Hardware Architectures for Post-Quantum Digital Signature Schemes

Deepraj Soni • Kanad Basu • Mohammed Nabeel
Najwa Aaraj • Marc Manzano • Ramesh Karri

Hardware Architectures for Post-Quantum Digital Signature Schemes

Springer

Deepraj Soni
NYU Tandon School of Engineering
New York, NY, USA

Mohammed Nabeel
Research Engineer at the Center for Cyber Security
NewYork University Abu Dhabi (CCS-NYUAD)
Abu Dhabi, United Arab Emirates

Marc Manzano
Executive Director of the Cryptography Research Centre
Technology Innovation Institute (TII)
Abu Dhabi, United Arab Emirates

Kanad Basu
Department of Computer Engineering
University of Texas
Dallas, TX, USA

Najwa Aaraj
Chief Research Officer
Technology Innovation Institute (TII)
Abu Dhabi, United Arab Emirates

Ramesh Karri
Electrical and Computer Engineering
New York University
Brooklyn, NY, USA

ISBN 978-3-030-57681-3 ISBN 978-3-030-57682-0 (eBook)
https://doi.org/10.1007/978-3-030-57682-0

This Springer imprint is published by the registered company Springer Nature Switzerland AG
The registered company address is: Gewerbestrasse 11, 6330 Cham, Switzerland

Foreword by Daniel Apon

Public-key cryptography was initiated in the mid-1970s by the revolutionary work of Whit Diffie and Martin Hellman. At that time, the development of the first cheap digital hardware freed the age-old art of cryptography from the limitations of hands-on mechanical computing and the physical distribution of sensitive messages. The low cost of deploying high-grade hardware rendered cryptography useful for commercial applications such as automated teller machines (ATMs) and personal computing devices. There was a new and inherent need for two parties to transmit information to one another inexpensively and effortlessly, but which would remain *hidden* to any third-party or eavesdropper. This trend accelerated in the 1990s with the wide-scale commercial deployment of the Internet and related communication technologies.

A fundamental goal in public-key cryptography is to establish an authenticated private channel between two parties over an untrusted communication medium. Typically, this is handled by two distinct technologies: a *key exchange mechanism* to establish private transmission of data and a *digital signature scheme* to authenticate communicating with the correct party. For nearly fifty years, we relied on the original idea of the Diffie–Hellman key exchange along with influential cryptosystems such as RSA (developed by Rivest, Shamir, and Adleman in the 1970s) and the related mathematical concepts to secure global telecommunications.

The security of these traditional public-key cryptosystems relies on the computational difficulty of factorizing the product of two large prime integers and finding discrete logarithms. Yet in 1994, Peter Shor demonstrated a breakthrough algorithm for a hypothetical type of computer—a *quantum computer*—that can quickly solve these challenging mathematical problems. Fortunately, for those deploying commercial applications of public-key cryptography over the Internet, building such a machine was out of the scope of engineering capabilities of the day. But in recent years, there has been a substantial amount of research on quantum computers. If large-scale quantum computers are ever built, they would gravely undermine the integrity and confidentiality of our current communications infrastructure on the Internet and elsewhere.

We stand today on the brink of a massive evolutionary step in modern communications: the imminent transition from traditional cryptography to the so-called *post-quantum cryptography*. The goal of post-quantum cryptography is to develop cryptographic systems that are secure against both quantum and classical computers and can inter-operate with existing communication protocols and networks. To facilitate this transition, the National Institute of Standards and Technology (NIST) initiated a worldwide process in late 2017 to solicit, evaluate, and standardize one or more quantum-resistant public-key cryptographic algorithms, during which 69 submissions of post-quantum key exchange mechanisms and digital signature schemes were received from around the world. In early 2019, this list of candidates was brought down to 26–17 of which are key exchange mechanisms and 9 of which are digital signature algorithms; the end of this second round of evaluations is imminent. It is expected that new, post-quantum cryptography standards to replace the traditional cryptosystems will be formally announced in 2022, and then large-scale deployment will commence.

A crucial goal during this transition period is the scientific examination and benchmarking of the performance of candidate-algorithms in various protocols, platforms, and types of hardware. In this book, the authors present the first comprehensive suite of experiments for hardware designs of NIST's post-quantum candidate-algorithms. The approach taken is the use of a High-Level Synthesis (HLS) tool to produce hardware implementations on FPGAs and ASICs. This is an exciting and important first step into the expansive world of post-quantum cryptographic hardware, which will drive global communications and commerce for future decades.

Furthermore, twenty years after NIST standardized the Advanced Encryption Standard (AES) for private-key communication, the serious scientific study of AES in hardware persists today, either to further accelerate the runtime of AES or to quest for side-channel attacks on the hardware running AES. Inevitably, a similar situation will occur in the case of post-quantum public-key cryptography over the next twenty years or so—both in software and hardware, across a diverse set of platforms, and within innumerable communication protocols yet to be designed.

Future research will undoubtedly benefit from this pioneering study in post-quantum hardware performed here. We highlight some future research directions in anticipation of the retrospective look at this present research activity and the insights drawn from this study. Other approaches for hardware design exist, such as the direct use of Verilog or VHDL, or implementing some form of hardware/software co-design—each with its respective pros and cons compared to the HLS approach employed here; these would be interesting to explore in depth as well. Furthermore, once highly optimized hardware designs are completed for the eventual finalists and future standardized algorithms, it will become possible to explore side-channel attacks against specific hardware implementations. A side-channel adversary may go beyond the typical notion of a passive eavesdropper and instead actively interfere with the operation of the cryptographic device to try and extract useful secret information. One example, among many, would be an adversary that injects faults into the hardware during the execution of a cryptographic algorithm (using a power surge or some other mechanism) in the hope that the device will mistakenly output information that reveals the secret key embedded in the device.

These issues, and many others, will need to be carefully considered by future engineers responsible for real-world deployment of post-quantum cryptography. It will not be an easy task. It will require the long-term, successful collaboration of mathematicians, computer scientists, cryptographers, hardware engineers, and software developers to make significant progress in this area. It is a crucial undertaking that needs to be done *now* rather than later: large-scale quantum computers are coming, perhaps sooner than we expect. One possible attack scenario would be an adversary who intercepts and stores communications sent over the course of several years in the traditional, pre-quantum cryptography era, waiting for the day that a quantum computer emerges so that they can use the accumulated data to retrospectively break the cryptosystem. The only defense against this scenario would be the rapid deployment of post-quantum cryptography, carefully studied by security analysts and benchmarked by engineering researchers to ensure its suitability for the commercial Internet and other communication platforms. The savvy researcher is warmly encouraged to join us on this difficult but exciting journey.

National Institute of Standards and Technology
Gaithersburg, MD, USA
May 14, 2020

Daniel Apon

Foreword by Stephen Neuendorffer

Algorithms are the core of encryption and security. Without good algorithms designed to be resilient to current and future cryptographic attacks, we cannot have long-term security. However, at a certain point, the realization of these algorithms also becomes important. While many algorithms can be succinctly captured as C programs and compiled for microprocessors, there are often important tradeoffs to be made. Some of these may be relatively simple, such as implementing simple functions using a table-lookup in memory instead of using a sequence of machine instructions. Other implementation tradeoffs around threading, cache hierarchy, and I/O constraints may be less obvious. This complexity only gets worse when we look to design specialized hardware accelerators for algorithms implemented in FPGAs and ASICs, where fine-grained concurrency and distributed memory allow a huge space for cost vs. performance optimization. The complexity of hardware design also comes with new languages, such as Verilog and VHDL, making it hard to migrate from working microprocessor code.

High-level synthesis tools, such as Xilinx Vivado HLS, blur the line between software programming and hardware design. Vivado HLS enables existing C/C++ code to be quickly implemented in Xilinx FPGAs, removing many of the mechanical aspects of hardware implementation. This enables designers to quickly identify cost or performance bottlenecks in a design and to begin exploring and optimizing the hardware implementation. This exploration is largely controlled through tool directives allowing fine-grained control of design tradeoffs in the context of overall design goals, such as clock frequency. As a result, designers can focus on complex architectural questions, such as resource sharing, data organization, and system integration, rather than the details of low-level hardware design.

This work represents a key first step to begin exploring the implementation space of new post-quantum cryptographic algorithms. By adapting the reference processor-oriented code to HLS-friendly code, the authors have demonstrated the feasibility of hardware implementations of these algorithms and provided a baseline to compare other hardware implementations too. The path is clear for a wider analysis of cost vs. performance tradeoffs and optimization of these implementations in concert with analysis of the algorithms themselves.

Xilinx Research Labs
San Jose, CA, USA
June 24, 2020

Stephen Neuendorffer

Preface

In this study, we developed and applied a uniform hardware design methodology and a common set of optimizations to implement and benchmark a subset of post-quantum cryptography (PQC) algorithms. Since the PQC algorithms are complex, it is challenging to develop functionally correct handcrafted, optimized register-transfer level (RTL) designs in Verilog/VHDL hardware description languages in a limited time. We adopted a high-level synthesis (HLS)-based design methodology starting off with the C specifications submitted to NIST evaluation.

This study is a collaboration between cryptographers from Technology Innovation Institute in Abu Dhabi and hardware designers from NYU. The cryptographers from Technology Innovation Institute in Abu Dhabi introduced the algorithmic aspects of the PQC algorithms and offered guidance to the design team. The NYU design team made the C-codes submitted to the NIST competition HLS-ready. Starting with these C specifications that are capable of synthesis by HLS tools, we conducted design-space explorations. This book is intended to benefit instructors and researchers interested in developing PQC hardware primitives. Instructors can use the HLS-synthesizable C-code repository mentioned in Table 1 to develop design and design-space exploration exercises for senior-level students in hardware modeling and design classes. Exercises and projects can focus on design-space exploration, high-level synthesis, HLS-based optimizations, and mapping the designs on a range of field programmable gate arrays (FPGA). The students will enhance their hardware design and modeling skill sets while becoming familiar with the next generation of crypto primitives.

Table 1 Links for HLS-ready C-code for the PQC algorithms

Algorithm	Link
CRYSTALS-Dilithium	https://github.com/deepraj88/CRYSTALS-Dilithium
Falcon	https://github.com/deepraj88/FALCON_Final
qTesla	https://github.com/deepraj88/qTESLA
LUOV	https://github.com/deepraj88/LUOV
MQDSS	https://github.com/deepraj88/MQDSS
Rainbow	https://github.com/deepraj88/Rainbow_Round2
Picnic	https://github.com/deepraj88/PICNIC
SPHINCS$^+$	https://github.com/deepraj88/SPHINCS
Official book web page—Latest hardware design updates	
NIST PQC round-2 signature schemes	https://wp.nyu.edu/hipqccheck/book-pqc-signature-schemes/

Researchers can use the developed HLS-ready C-codes for PQC algorithms and the design framework as a starting point for developing more optimized hardware implementations. Furthermore, they can perform side-channel and fault-attack resistance analysis on these accelerators and make changes to the specifications and the designs to make them side-channel resistant.

New York, NY, USA — Deepraj Soni

Dallas, TX, USA — Kanad Basu

Abu Dhabi, United Arab Emirates — Mohammed Nabeel

Abu Dhabi, United Arab Emirates — Marc Manzano

Abu Dhabi, United Arab Emirates — Najwa Aaraj

Brooklyn, NY, USA — Ramesh Karri

June 1, 2020

Acknowledgments

The research was supported by the National Science Foundation under grants CNS-1526405 and CNS-1513130, Office of Naval Research under grant N00014-18-1-2672, and NYU and NYU Abu Dhabi, Center for Cyber Security (CCS and CCS-AD). The contents of this book do not reflect the opinion of the United States Government or any other funding agencies.

The authors would like to thank their family and friends for continued support while writing this book. We are grateful to the following people, who helped us in various stages of this project:

- Xilinx team for their instructions on HLS tool operations and Xilinx University Program (XUP) for providing the licenses.
- Dr. Nina Bindel for her support in resolving qTESLA C-code issues.
- Ward Beullens for his support in resolving LUOV C-code issues.
- Dr. Greg Zaverucha and Dr. Sebastian Ramacher for their support in resolving Picnic C-code issues.
- Dr. Jintai Ding for his support in resolving RAINBOW C-code issues.
- Dr. Ludovic Perret for his support in resolving GeMSS C-code issues.
- Dr. Andreas Huelsing for his support in resolving SPHINCS$^+$ hash function issues.
- Mr. Kalpan S. Mehta (NYU) for making the Rainbow C-code HLS-ready.
- Dr. Victor Mateu (Technology Innovation Institute) for supporting the analysis of the PQC schemes covered in the book.
- Santos Merino del Pozo (Technology Innovation Institute) for providing feedback and supporting this research.
- Philip Rodenbough and the NYUAD Scientific Writing Program for developmental manuscript feedback.

Finally, we applaud and appreciate the efforts of everyone involved in developing the post-quantum cryptography algorithms.

Contents

About the Authors

Deepraj Soni is a Ph.D. student at NYU Tandon School of Engineering. Deepraj works on hardware implementation, evaluation, and security of post-quantum cryptographic algorithms. He received his M.Tech from the Department of Electrical Engineering, Indian Institute of Technology Bombay (IIT-B). His thesis focused on developing a framework for hardware–software co-simulator and neural network implementation on an FPGA. After graduation, Deepraj worked as a design engineer in the semiconductor division of Samsung and SanDisk. At Samsung, he was responsible for the design and architecture of the image processing IPs such as region segmentation and Embedded CODEC. He was also responsible for communication IPs such as FFT/IFFT, time & frequency deinterleaving, and demapper for canceling the noise. At SanDisk, Deepraj helped in the development of system-on-chip (SoC) level design for the memory controller.

Kanad Basu received his Ph.D. from the Department of Computer and Information Science and Engineering, University of Florida. His thesis was focused on improving signal observability for post-silicon validation. Post-PhD, Kanad worked in various semiconductor companies like IBM and Synopsys. During his Ph.D., Kanad interned at Intel. Currently, Kanad is an Assistant Professor at the Electrical and Computer Engineering Department of the University of Texas in Dallas. Prior to this, Kanad was an Assistant Research Professor at the Electrical and Computer Engineering Department of NYU. He has authored 2 US patents, 2 book chapters, and several peer-reviewed journal and conference articles. Kanad was awarded the "Best Paper Award" at the International Conference on VLSI Design 2011. Kanad's current research interests are hardware and systems security. Kanad is an Associate Editor for IET Computers and Digital Technology Journal and a Guest Editor for Springer Journal of Electronic Testing. Kanad has served as a Program Committee member for various conferences including the Design Automation Conference, VLSI Design Conference, and Asian Test Symposium, among others. In addition, he has arranged special sessions at conferences like the VLSI Test Symposium and participated in industry panels on machine learning.

Mohammed Nabeel received his B.Tech degree in electrical and electronics engineering from National Institute of Technology Calicut (NIT-C), India. He is currently working as a Research Engineer at the Center for Cyber Security at New York University Abu Dhabi (CCS-NYUAD). Apart from working on research in the field of hardware security, his job also includes implementing and prototyping the research ideas in ASIC and FPGA. He has around 12 years of industry experience in chip design specializing in microarchitecture, protocol know-how, RTL design, functional verification, FPGA prototyping, synthesis, static timing analysis, and post-silicon bring up. Before joining CCS-NYUAD, he worked at Texas Instruments as a Staff Engineer and at Qualcomm as Senior Engineer Lead. He has more than ten conference and journal papers along with two issued US patents.

Dr. Najwa Aaraj is the Chief Research Officer at the Technology Innovation Institute (TII), which is based in the United Arab Emirates (UAE). In her role, she leads the research and development of cryptographic and quantum communication technologies, including post-quantum cryptography (PQC) software libraries and hardware implementations, lightweight cryptographic libraries for embedded and RF systems, cryptanalysis, quantum key distribution, quantum random number generation, and applied machine learning for cryptographic technologies. Dr. Aaraj earned a Ph.D. with highest distinction in Applied Cryptography and Embedded Systems Security from Princeton University (USA). She has extensive expertise in applied cryptography, trusted platforms, security architecture for embedded systems, software exploit detection and prevention systems, and biometrics. She has over 15 years of experience with global firms, working in multiple regions ranging from Australia to the USA. Before joining TII, Dr. Aaraj was the Senior Vice President of Products & Cryptography Development at DarkMatter, now part of Digital14, a cybersecurity leader based in the UAE. She was formerly at Booz & Company, where she led consulting engagements in the communication and technology industry for clients globally. She also held a Research Fellow position with the Embedded Systems Security Group at IBM T.J. Watson Security Research in New York State, and with the Intel Security Research Group in Portland, Oregon, where she worked on Trusted Platform Modules and contributed to an early prototype of a Trusted Platform Module (TPM) 2.0 based firmware. She was also a Research Staff Member at the NEC Laboratories in Princeton, New Jersey. Dr. Aaraj has written multiple conference papers, IEEE and ACM journal papers, and book chapters. She has also received patents on applied cryptography, embedded system security, and machine learning-based protection of IoT systems.

Dr. Marc Manzano is the Executive Director of the Cryptography Research Centre at the Technology Innovation Institute (TII), a UAE-based scientific research center. In this role, he supervises and coordinates the R&D teams, spearheads the definition of research roadmaps, is responsible for the design, implementation, and quality of both software and hardware cryptographic libraries, and oversees the establishment of strategic partnerships with international universities and research institutes. His current research interests include post-quantum cryptography, lightweight cryptography, the intersection between machine learning and cryptanalysis, performance optimizations of cryptographic implementations on a wide range of architectures, and quantum algorithms. He has presented more than 25 articles at international conferences, published more than ten journal papers, and collaborated on several scientific books related to cryptography and computer network security. Over the past ten years, Dr Manzano has led the development of many secure cryptographic libraries and protocols. Dr. Manzano was formerly the Executive Director of the Cryptography Research and Development division at DarkMatter, now part of Digital14. Prior to that, he held a position at Scytl where he was responsible for implementing pivotal cryptographic components of an electronic voting platform. Furthermore, Dr. Manzano also worked at Marfeel, where he contributed to the creation of a data collection and prediction engine utilizing machine learning techniques. Dr. Manzano holds a Ph.D. in Computer Network Security, which he earned under the supervision of the University of Girona (Spain) and Kansas State University (the USA). He earned an MSc in Computer Science from the University of Girona (Spain) while doing research stays at UC3M (Spain) and DTU (Denmark). He started off on his research career while finalizing his BSc in Computer Engineering at Strathclyde University (UK).

Ramesh Karri is a Professor of Electrical and Computer Engineering at New York University. He co-directs the NYU Center for Cyber Security (http://cyber.nyu.edu). He also leads the Cyber Security Trust of the NY State Center for Advanced Telecommunications Technologies at NYU. He co-founded the Trust-Hub (http://trust-hub.org) and organizes the Embedded Systems Challenge (https://csaw.engineering.nyu.edu/esc), the annual red team blue team event. Ramesh Karri has a Ph.D. in Computer Science and Engineering, from the University of California San Diego and a B.E in ECE from Andhra University. His research and education activities in hardware cybersecurity include trustworthy integrated circuits; processors and cyber-physical systems; security-aware computer-aided design, test, verification, validation, and reliability; nano meets security; hardware security competitions, benchmarks, and metrics; biochip security; additive manufacturing security. He has published over 280 articles in leading journals and conference proceedings. Karri's work on hardware cybersecurity received best paper nominations (in ICCD 2015 and DFTS 2015) and awards (in ACM TODAES 2018, ITC 2014, CCS 2013, DFTS 2013, and VLSI Design 2012). He received the Humboldt Fellowship and the National Science Foundation CAREER Award. He serves(d) on the editorial boards of several IEEE and ACM Transactions (TIFS, TCAD, TODAES, ESL, D&T, and JETC). He is a Fellow of the IEEE and served as an IEEE Computer Society Distinguished Visitor in the years 2013–2015. He served on the Executive Committee of the IEEE/ACM Design Automation Conference leading the SecurityDAC initiative (2014–2017). He delivered invited keynotes, talks, and tutorials on Hardware Security and Trust (ESRF, DAC, DATE, VTS, ITC, ICCD,

NATW, LATW, CROSSING, etc.). He co-founded the IEEE/ACM NANOARCH Symposium and served as the program/general chair of several conferences (IEEE ICCD, IEEE HOST, IEEE DFTS, NANOARCH, RFIDSEC, and WISEC). He serves on several program committees, including DAC, ICCAD, HOST, ITC, VTS, ETS, ICCD, DTIS, and WIFS.

Acronyms

AES	Advance Encryption Standard
ASIC	Application-specific integrated circuit
BRAM	Block random access memory
CMA	Chosen message attack
CRYSTALS	Cryptographic Suite for Algebraic Lattices
DES	Data Encryption Standard
DSP	Digital signal processor
EDA	Electronic Design Automation
EUF-CMA	Existential unforgeability under chosen message attack
FALCON	Fast Fourier lattice-based compact signatures over NTRU
FF	Flip-flop
FPGA	Field programmable gate array
FORS	Forest of random subset
HDL	Hardware description language
HLS	High-level synthesis
KAT	Known answer test
KEM	Key encapsulation mechanism
LUT	Look-up table
LPE	Low-power enhanced
LWE	Learning with errors
MPC	Multi-party computation
NTRU	Nth degree truncated polynomial ring
NIST	National Institute of Standards and Technology
NTT	Number theory transform
PKE	Public-key encryption
PQC	Post-quantum cryptography
PRF	Pseudo random function
PRG	Pseudo random generator
QROM	Quantum random Oracle model
RAM	Random access memory
ROM	Read-only memory

R-LWE	Ring learning with errors
RTL	Register-transfer level
SHA	Secure hash algorithms
SHAKE	Secure hash algorithm Keccak
SIS	Short integer solution
SRAM	Static random access memory
SUF-CMA	Strong unforgeability under chosen message attacks
VHDL (VHSIC-HDL)	Very high speed integrated circuit hardware description language
VLSI	Very large-scale integration
WOTS	Winternitz one-time signature
XKCP	eXtended Keccak Code Package

Chapter 1
Introduction

1.1 Post Quantum Cryptography (PQC)

Substantial advances in quantum computing in the past decade have reassured the scientific community about the necessity to build quantum-resistant cryptosystems [1]. Post-Quantum Cryptography (PQC) has emerged as the preferred solution to face the threats that quantum computers pose to traditional public-key cryptography based on number theory [2] (i.e., integer factorization or discrete logarithms). Lattice-based cryptography, multivariate cryptography, hash-based cryptography schemes, isogeny-based cryptography, and code-based cryptography can be used to design cryptosystems, which are secure against attacks launched on classical computers and potentially quantum computers [3]. Thus, these strategies are regarded as PQC algorithms.

Lattice-based cryptography algorithms offer simple and efficient implementation, with strong proof of security [4]. Lattice-based cryptography builds on the hardness of the shortest vector problem (SVP), which entails approximating the minimal Euclidean length of a lattice vector for arbitrary basis. To solve SVP, the worst-case quantum polynomial time is around $\exp(\widetilde{O}(\sqrt{n}))$. Even with a quantum computer's computational power, the complexity of SVP is polynomial in n [5]. Short Integer Solutions (SIS) is one of the many problems that is part of lattice family. SIS problems are secure in the average case if the SVP is hard in the worst case[6].

Code-based cryptography schemes have underlying assumptions that generator matrix and random matrix are indistinguishable and generic decoding is hard. These schemes follow a conservative approach for public key encryption/key encapsulation, as it is based on a well-studied problem [7]. This class of algorithms becomes vulnerable if large key size is reduced [8]. Researchers have recommended approaches to reduce the key size without security compromise [9, 10].

Multivariate cryptography builds on the difficulty of solving the finite field multivariate polynomial (MVP) problem. Solutions of MVP problems are NP-hard

D. Soni et al., *Hardware Architectures for Post-Quantum Digital Signature Schemes*, https://doi.org/10.1007/978-3-030-57682-0_1

over any field. If all equations are quadratic over $GF(2)$, MVPs are NP-complete problems [11]. Though some MVP-based schemes have been shown to be vulnerable [12], it allows competitive signature sizes for PQC signature scheme.

Hash-based digital signatures are based on the security properties of the underlying symmetric primitives, more specifically cryptographic hash functions (leveraging properties of collision resistance and second pre-image resistance).

In 2016, the National Institute of Standards and Technology (NIST) announced its intention to start a standardization effort to define quantum resistant standards for Key Encapsulation Mechanism (KEM) and Public Key Encryption (PKE), and digital signatures [5]. In the call for proposals, NIST defined five different security strengths directly related to NIST standards in symmetric cryptography:

- Security Level 1: Algorithm is at least as hard to break as AES128 (weak in terms of quantum resistance—Exhaustive Key Search).
- Security Level 2: Algorithm is at least as hard to break as SHA256 (strong in terms of quantum resistance—Collision Search).
- Security Level 3: Algorithm is at least as hard to break as AES192 (stronger in terms of quantum resistance—Exhaustive Key Search).
- Security Level 4: Algorithm is at least as hard to break as SHA384 (very strong in terms of quantum resistance—Collision Search).
- Security Level 5: Algorithm is at least as hard to break as AES256 (strongest in terms of quantum resistance—Exhaustive Key Search).

The first round of the NIST PQC Competition started in December 2017 and received 82 submissions, from which 69 were accepted: 19 digital signature candidates and 45 KEM/PKE schemes [13]. In January 2019, the second round candidates of the NIST PQC Competition were announced [14]: 9 digital signature candidates and 17 KEM/PQC schemes [15]. NIST has publicly announced a third round starting around June 2020[16], just as the present work is going to press. Round-2 candidates, corresponding scheme, and NIST security level mapping are summarized in Table 1.1.

1.2 Book Roadmap

This book presents a hardware assessment of eight digital signatures candidates from Round-2 of the NIST PQC Competition. Chapters 2, 3, and 4 explore lattice-based algorithms CRYSTALS-Dilithium, Falcon, and qTESLA, respectively. Chapters 5, 6, and 7 explore multivariate cryptographic algorithms LUOV, MQDSS, and rainbow, respectively. Chapters 8 and 9 explore hash-based cryptographic schemes Picnic and SPHINCS$^+$, respectively. Subsequent sections of this chapter explain the FPGA and ASIC hardware design flows.

Table 1.1 Summary of Round-2 candidates of the NIST PQC competition

Algorithm	Scheme	Type	NIST security level				
			1	2	3	4	5
BIKE	Code	PKE/KEM	X		X		X
Classic McEliece	Code	PKE/KEM	X		X		X
CRYSTALS-KYBER	Lattice	PKE/KEM	X		X		X
FrodoKEM	Lattice	PKE/KEM	X		X		X
HQC	Code	PKE/KEM	X		X		X
LAC	Lattice	PKE/KEM	X		X		X
LEDAcrypt	Code	PKE/KEM	X		X		X
NewHope	Lattice	PKE/KEM	X				X
NTRU	Lattice	PKE/KEM	X		X		X
NTRU Prime	Lattice	PKE/KEM		X	X	X	
NTS-KEM	Code	PKE/KEM	X		X		X
ROLLO	Code	PKE/KEM	X		X		X
Round5	Lattice	PKE/KEM	X		X		X
RQC	Code	PKE/KEM	X		X		X
SABER	Lattice	PKE/KEM	X		X		X
SIKE	Isogeny	PKE/KEM	X	X	X		X
Three Bears	Lattice	PKE/KEM		X		X	X
CRYSTALS-Dilithium	Lattice	Digital signature	X	X	X		
FALCON[a]	Lattice	Digital signature	X	X	X	X	X
FALCON[b]	Lattice	Digital signature	X			X	X
GeMSS	Multivariate	Digital signature	X		X		X
LUOV	Multivariate	Digital signature	X		X		X
MQDSS	Multivariate	Digital signature	X	X	X	X	
Picnic	Hash-based	Digital signature	X		X		X
qTESLA[c]	Lattice	Digital signature	X	X	X		X
qTESLA[b]	Lattice	Digital signature	X		X		
Rainbow	Multivariate	Digital signature	X		X		X
SPHINCS$^+$	Hash-based	Digital signature	X		X		X

[a] As of the beginning of NIST round 1 PQC Competition—Considered in this book
[b] As of April 2020
[c] As of the beginning of NIST round 2 PQC Competition—Considered in this book

1.3 Hardware Design Using High-Level Synthesis

VLSI technology has reached a transistor count density of over one billion for microprocessors [17, 18]. However, the system performance has not been able to abide by Moore's law; thus, motivating designers to explore nontraditional design alternatives [19]. Traditional design approaches use Hardware Description Language (HDL)-based specification of an algorithm and proceed through design steps like verification, logic synthesis, placement and routing, timing analysis, testing, and validation to produce the final manufactured Integrated Circuit (IC). This process incurs considerable timing overhead. HDL languages, such as Verilog and VHDL, describe digital logic with register transfer level (RTL) without specifying an underlying technology (CMOS) or a technology node (65 nm or 45 nm). With increasing complexity of the chip, HDL-based design development becomes challenging and time consuming. Hence, a design tool to work on a more abstract level is necessary. High-Level Synthesis (HLS) generates RTL hardware descriptions from high-level language (e.g., C/C++) specifications.

The semiconductor industry is continually refining the design flow to enable a fast development cycle to increase designer productivity, and to reduce time-to-market of new innovations. Design at a high level and automation of the design process have become integral parts of all semiconductor design centers. Many high-level synthesis tools, such as Vivado HLS [20], LegUp [21], CyberWorkbench [22], Catapault [23], Stratus [24], and Bluespec [25], have been developed in recent years.

The RTL-based design approach can provide designs with optimal area overhead or latency. However, it requires a substantial amount of time by expert design engineers to build a complex hardware. RTL architecture is not flexible because of its preference for structured code. Change in architecture requires repeating the design and verification process, which significantly impacts the design time and the time to market schedules. On the other hand, the emerging HLS-based approach enables faster and easier implementation, since it rapidly produces different RTL from the same high-level description, without the need for hand-stitching by design engineers. Though HLS-generated designs may not always be optimized, they provide a unique advantage of modifying the RTL in a short time window to obtain different RTL variants of the same behavioral specification. This flexibility in rapidly exploring the design space of the hardware implementation of an algorithm and in understanding the trade-offs between different designs of the same algorithm is essential in improving designer productivity. Moreover, high-level functional specifications are easier to understand and change in at the high level in a language such as C as compared to a low-level hardware description language (HDL) such as VHDL and Verilog.

Figure 1.1 shows the HLS-based design flow. The input to any HLS tool is a high-level language (HLL) specification of an algorithm. Depending on the tool, HLL specification can be in C, C++, SystemC, C#, or MATLAB. Apart from this, the tool requires a technology library to construct a design that is technology-aware and a constraint file to specify algorithm-specific optimizations. Using a state-of-the-art compiler, HLS analyzes high-level language specification and performs optimization to reduce complexity of the input algorithm. Transformation phase converts the higher level input code to intermediate representation, e.g., LLVM [26] and proprietary control data flow graph (CDFG) representations. Scheduling maps operations to the smallest number of control steps subject to a given cost constraint or to minimize the cost subject to a given number of control steps. Resource binding maps the data path and control operations to functional units, storage units, and interconnection units, while keeping the design behavior intact. The synthesis step uses all the intermediate representation and analysis to create the HDL model. The

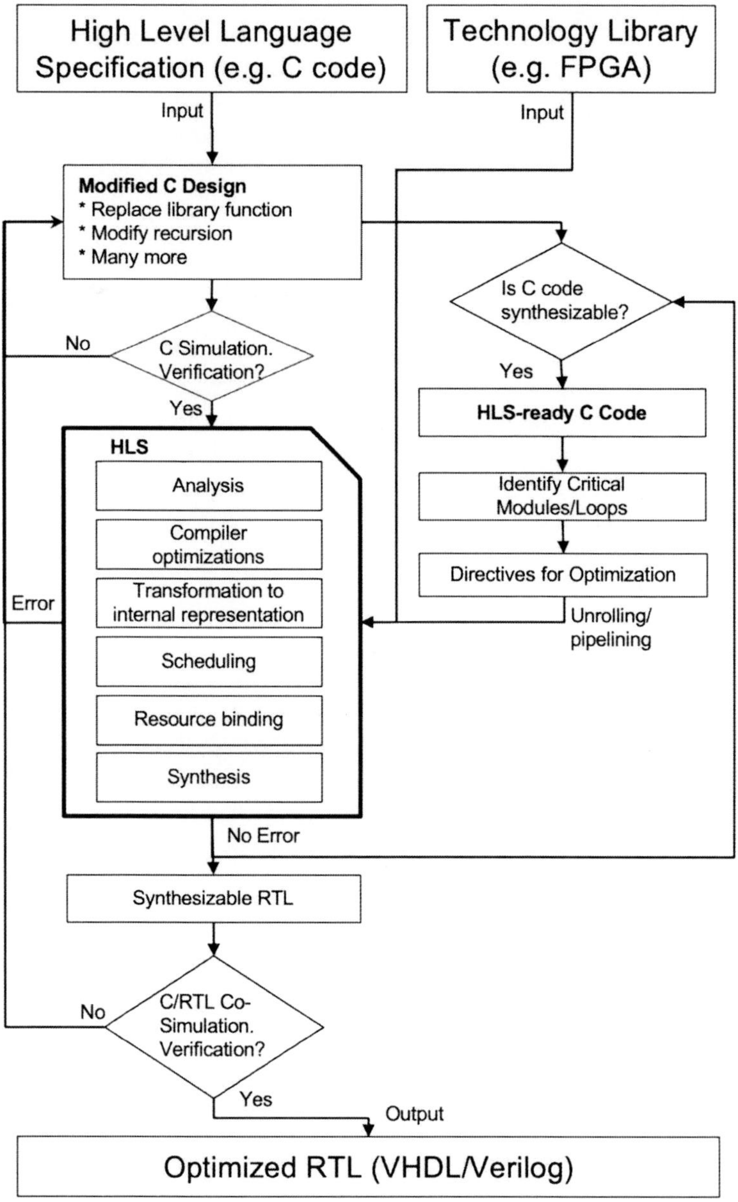

Fig. 1.1 C-to-FPGA design flow: NIST submission C code → HLS-synthesizable C code → RTL

primary output of any HLS tool is the RTL model of the hardware accelerator. Besides this, the tool provides a testbench used to verify the generated RTL model.

For all HLS designs in this book, we use Xilinx Vivado HLS 2018.2. C-based description of the PQC algorithms are used as input. NIST has recommended to map the PQC hardware implementations onto the Artix-7 FPGA for its small size and energy efficiency. Hence, Artix-7 XC7A200T-2FBG676C FPGA is a target device for all the designs. We use several HLS directives to improve the area overhead and latency of the design. Using these inputs and constraints, Vivado HLS generates synthesizable RTL and a testbench, either in Verilog or in VHDL.

While compiling and executing a C program, a software developer has the ability to fix various errors including syntax errors, compilation errors, and functional errors, in order to ensure the program executes correctly. However, such a functionally correct C code might not be synthesizable by HLS into hardware. HLS has its own set of rules guided by the custom hardware that it targets, which the C code needs to abide by. For example, HLS does not understand dynamic allocation or arrays. This is because, the amount of memory in hardware is fixed. Therefore, an input C program requires additional effort to make it synthesizable into RTL components. We developed the design flow, which includes modifying the C code in such a way that HLS can compile, synthesize, and generate the RTL model. This book refers to this C code as *HLS-ready C code*.

With HLS-based design flow, an alternate path for program translation has emerged which translates C code in a gate-level netlist. Simplicity and the behavioral nature of high-level languages make the design specifications at this level more intuitive, thus bridging the gap between hardware and software. In the next section we describe the HLS-based design flow for Field Programmable Gate Arrays (FPGAs). The last section of this chapter details the HLS-based design flow for the Application Specific Integrated Circuits (ASIC).

1.3.1 FPGA Design Flow

In this section, we will analyze the HLS FPGA design flow. Figure 1.1 shows the HLS FPGA design flow for PQC algorithms. This is a generic flow where the high level description of the algorithm is in C. We divide this design flow into two parts: preparing HLS-ready C code and generating optimized RTL. The first part of the flow modifies the C code. This part produces HLS-ready C code and the HLS-generated RTL. However, the design generated from this RTL is non-optimized and has higher area overhead and latency. To improve the area and latency of the design, the second part of the flow analyzes the code, finds modules for optimization, and generates an optimized design.

Figure 1.1 shows that the flow modifies the original C code to make it synthesizable by the HLS tool. For each NIST Round-2 PQC candidate, we use the C specification as input to the HLS-based design flow. We: (i) replace library functions, (ii) change complex hierarchy of structure, (iii) change variable length arrays and pointers to a fixed-dimension array, (iv) remove file operations, (v) replace recursions, and (vi) modify complex pointer operations to make the C code

HLS-synthesizable. These changes in the C code expose the difference between the high-level software implementations targeting general-purpose processors and application-specific hardware implementations.

We replace library function calls in the C language with equivalent high level C functions. High-level C functions might use an optimized assembly language code snippet. Such assembly language snippets cannot be understood and processed by HLS. Therefore, they need to be replaced with equivalent C code. Library functions implement optimized C solutions which, in turn, require complex operations or more resources in hardware. Accordingly, we replace `memcpy`, `memcmp`, `memmove`, `randombytes`, and other library functions with equivalent C-level functions, that we developed. After each of these replacements, C simulation is performed to verify that the modification does not alter the system functionality. HLS tools are not efficient with file operations and they do not synthesize complex pointer operations. To resolve this issue, we change the C code.

Structures in C are typically a collection of different data types that can produce a faster design, capture the semantics accurately, and facilitate the implementation quickly. HLS can understand the simple structures in C with a few variables or fixed-size arrays. Complex constructs, such as structures within a structure or pointers inside a structure, are not supported by the current generation of HLS tools. To make such structures synthesizable, we convert pointers to fixed-size arrays in structures and break the complex structure into separate data types.

Dynamic memory allocation creates or destroys flexible storage when corresponding function calls are executed. At run-time, an operating system running on a processor typically allocates memory for the program to use. This is based on the idea of virtual memory for each program which is mapped to the fixed memory at run-time. However, hardware accelerators generated by a HLS tool do not dynamically map a larger memory request onto a smaller, predefined fixed-size Random Access Memory (RAM) or Read-Only Memory (ROM). That is, HLS does not support dynamic memory allocation. Hence, we replace all dynamic memory allocations with static memory allocations, i.e., with fixed-size arrays.

While pointers are helpful in low-level operations, the corresponding C language constructs are complex to map into hardware. In fact, while programmers grapple with pointers, recent programming languages such as Java no longer support pointers. Because of the complexity and flexibility of pointer handling, HLS tools face difficulties in interpreting the correct functionality of such pointers. While the current generation of HLS tools (including Xilinx Vivado) support a handful of simple pointer operations, they do not support complex pointer operations and typecasting. Hence, to make the C code synthesizable, we substitute complex pointer handling with equivalent functionality without pointers.

File operations facilitate easy read/write access to data. However, they are not required for hardware accelerators, in particular to fetch and store data directly from memory. As such, we remove or modify all file operations. Current generation of HLS tools can only synthesize the C code after all the appropriate modifications are made.

After each change, we compare the results of the simulation of the modified C code with the Known Answer Tests (KAT) submitted to NIST. Once the C code is thoroughly verified and validated, we execute HLS on this C code. This way we ensure that HLS only runs on a functionally verified C code. Anytime HLS produces errors or warnings, we identify these non-synthesizable parts, change the code, and verify the change before running the HLS tool once again. After multiple iterations, the C→Synthesizable C conversion converges and the Vivado HLS is able to generate synthesizable RTL and a corresponding testbench.

HLS does not guarantee that the output RTL is functionally correct, even if the input C model is functionally verified. Hence, the designer needs to verify the output RTL. We use the Vivado HLS C-RTL co-simulation for this purpose. The C-RTL co-simulation uses both C and RTL code to verify the synthesizable RTL. Vivado HLS simulates the C code on KATs and collects all the input and output data of the top function. Vivado HLS ports the synthesizable RTL and the testbench to Vivado Integrated Synthesis Environment (ISE). Vivado runs the simulation whereby the top module is the testbench. The testbench furnishes input data collected from the C simulation in Vivado HLS, to the synthesizable RTL. Vivado runs a cycle-accurate simulation of synthesizable RTL. The final output generated by the Vivado simulation is compared to the output data collected from the top function by Vivado HLS. If the results of the C simulation in Vivado HLS and the RTL simulation in Vivado are identical, we claim that the synthesizable RTL is validated using KATs. This process does not deliver formal verification or complete functional coverage; it is only validated against the KATs provided by each of the teams in their NIST Round-2 PQC submission. To reiterate, the modified C code, which provides the verified synthesizable RTL, is *HLS-ready*.

The second part of the hardware design flow focuses on hardware analysis and optimization. Traditionally, hardware analysis and optimizations depend on the trade-off between performance and area [27]. Designs with inherent parallelism provide better performance at the expense of higher area overhead. Design and optimization constraints are chosen based on the target application requirements. In this book, we consider two additional dimensions of design-space exploration: security and power consumption. Since we are exploring cryptographic accelerators, the security parameter is crucial. We identify critical modules, loops, and functions in the HLS-synthesizable C code that incur either more latency or a larger area. We applied two performance optimizations on the critical modules: loop unrolling and loop pipelining. Different optimizations and constraints provide different optimized RTL versions, each of which has a unique architecture with distinct power, area, and speed metrics. The RTL with no performance optimization is the baseline implementation. The optimized RTL is used as input to the ASIC design flow.

After the HLS-synthesizable code is finalized, there are two major steps for the optimization of FPGA design flow: (i) identification of critical modules and (ii) optimization of the critical modules. The first step involves identification of critical functions that incur higher area or latency overhead. Optimizations made to these critical functions improve the area and latency numbers significantly. Design space exploration provides the range of design trade-off points. Based on requirements, designers select one optimal point: for example, if the security, power, and area are more important than latency (for low-powered IoT edge devices), a design with low area and high security is selected, irrespective of the latency.

1.3.2 ASIC Synthesis Flow

Once the RTL is coded for a design, ASIC Synthesis is performed to transform the RTL code into logic gates, returning a gate-level netlist [28, 29]. The gate-level circuit generated is logically optimized to meet the goals defined by user constraints. Typical constraints include: (i) target frequency, (i) area budget, and (ii) power profile. This process is typically automated using ASIC EDA (Electronic Design and Automation) tools. Besides the RTL code and a set of constraints, the tool requires a standard cell library. The standard cell library is a collection of low-level digital logic functions, such as logic gates and memory elements like flip flops and latches. After synthesis, design costs can be obtained, for metrics such as performance (i.e., the maximum frequency the design/netlist can support), power, and area (referred to as PPA) of the design/netlist from the synthesis tool. Based on post-synthesis reports, one can also identify performance bottlenecks and improve the input RTL/algorithms accordingly (Fig. 1.2).

The standard cell library is technology dependent. Synthesis of RTL code generated from HLS is done using 65 nm (nanometer) standard cell library from GlobalFoundries. The EDA tool used is Synopsys Design Compiler (DC). Following common practices, the standard cell library used for synthesis was the one characterized for the worst voltage (1.08V), temperature (125 °C), resistance, and capacitance. Synthesizing with such a library ensures that we can achieve the target frequency at the operational voltage of 1.2V±10% and at any temperature less than 125°C.

Clock Constraint The longest path in the design (i.e., critical path) determines the frequency at which the design can be run. To find the maximum frequency, designers

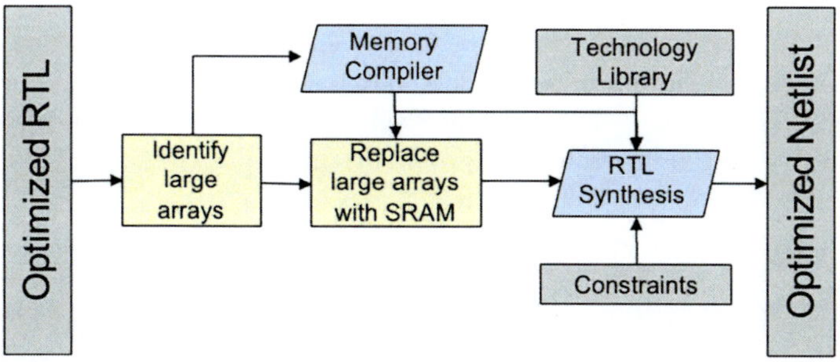

Fig. 1.2 C-to-ASIC design flow starts with the Optimized RTL and follows the Optimized RTL → memory map → Optimized Netlist design flow

needs to identify the clock constraint where the critical path starts failing to meet the target frequency. The clock constraint given to all the designs in this book is 200 MHz.

Memory The RTL code generated by the HLS implements the memory using Flip-Flops. On-chip memory can be synthesized using D Flip Flop (DFF), but synthesizing large memories can significantly impact the area, power, and delay of the overall design. Standard practice is to use Static Random Access Memory(SRAM)-based memory [30] to implement large memories. DFF-based memory, above 512 bits, is replaced with the SRAM equivalent. Cost-effective DFF-based memory is used for smaller memory requirements. SRAM is the best choice for larger memory requirements in terms of area, power, and delay. For example, DFF-based memory of qTESLA signature generation occupied 2.75 mm^2, whereas SRAM-based memory of qTesla Signature Generation occupied 1.31 mm^2, providing a 52.3% reduction of the area dimension. To generate SRAM models needed for the ASIC design flow, we use the ARM Artisan memory compiler [31]. The ARM memory compiler generates library models for the static RAM cells based on the configuration supplied by the designer. The ARM memory compiler supports only single-port RAMs (SPRAM), dual-port RAMs (DPRAM), and single-port Read-Only Memory (SPROM). Multi-port RAMs and ROMs are implemented with multiple instances of these supported memories. Similarly, for memories larger than the memory size supported by the compiler, we use multiple instances of the supported size.

Power The power consumed by a design depends on the switching activity of its nets—the wire connecting two logic cells in the netlist—and its cell pins. There are many ways of estimating the switching activity of the design for the specific computing power consumed. We use the method where the static probability (fraction of time the design object is at logic 1) and the toggle rate (the rate at which the design object switches between logic 0 and logic 1) of the design's input ports are used as input for power calculation. When computing the power consumption, it is assumed that input toggles every 10 clock cycles and the static probability is 50%. This is pessimistic in reality, but as the intent is to calculate the relative power with respect to other designs, the same pessimistic constraint is used for all designs.

Once the synthesis results are ready, the Power, Performance, and Area (PPA) are analyzed for all the optimization variants (baseline, loop unrolled, and pipelined) of all the algorithms. Apart from PPA, an analysis is carried out for parameters like the size of memory implemented using SRAM and flip flops, and the total flip flop counts for the different optimization techniques.

References

1. M.H. Devoret, R.J. Schoelkopf, Superconducting circuits for quantum information: An outlook. Science **339**(6124), 1169–1174 (2013)

2. P.W. Shor, Polynomial-time algorithms for prime factorization and discrete logarithms on a quantum computer. SIAM J. Comput. **26**(5), 1484–1509 (1997)
3. D.J. Bernstein, J. Buchmann, E. Dahmen, *Post Quantum Cryptography*, 1st edn. (Springer Publishing Company, 2008)
4. D. Micciancio, Lattice-based cryptography. Encycl. Cryptogr. Secur., 713–715 (2011)
5. L. Chen, S. Jordan, Y.-K. Liu, D. Moody, R. Peralta, R. Perlner, D. SmithTone, Report on post-quantum cryptography (2016)
6. R. Cramer, L. Ducas, B. Wesolowski, Short stickelberger class relations and application to ideal-svp, in *Annual International Conference on the Theory and Applications of Cryptographic Techniques* (Springer, 2017), pp. 324–348
7. R.J. McEliece, A public-key cryptosystem based on algebraic. Coding Thv **4244**, 114–116 (1978)
8. D.J. Bernstein, T. Lange, C. Peters, Attacking and defending the mceliece cryptosystem, in *International Workshop on Post-Quantum Cryptography* (Springer, 2008), pp. 31–46
9. H.A. Shehhi, E. Bellini, F. Borba, F. Caullery, M. Manzano, V. Mateu, An ind-cca-secure code-based encryption scheme using rank metric, in *Progress in Cryptology AFRICACRYPT 2019 11th International Conference on Cryptology in Africa, Rabat, Morocco, July 9-11, 2019, Proceedings*. Lecture Notes in Computer Science, vol. 11627 (Springer, 2019), pp. 79–96
10. P. Loidreau, A new rank metric codes based encryption scheme, in *International Workshop on Post-Quantum Cryptography* (Springer, 2017), pp. 3–17
11. N. Courtois, A. Klimov, J. Patarin, A. Shamir, Efficient algorithms for solving overdefined systems of multivariate polynomial equations, in *International Conference on the Theory and Applications of Cryptographic Techniques* (Springer, 2000), pp. 392–407
12. V. Dubois, P.-A. Fouque, A. Shamir, J. Stern, Practical cryptanalysis of sflash, in *Annual International Cryptology Conference* (Springer, 2007), pp. 1–12
13. NIST, Round 1 submissions (2018). Available at https://csrc.nist.gov/Projects/Post-Quantum-Cryptography/Round-1-Submissions
14. G. Alagic, J. Alperin-Sheriff, D. Apon, D. Cooper, Q. Dang, Y.-K. Liu, C. Miller, D. Moody, R. Peralta, R. Perlner, A. Robinson, D. SmithTone, Status report on the first round of the nist post-quantum cryptography standardization process (2019)
15. NIST, Round 2 submissions (2019). Available at https://csrc.nist.gov/Projects/Post-Quantum-Cryptography/Round-2-Submissions
16. NIST, Pqc forum e-mail list (2019). Available at https://csrc.nist.gov/Projects/Post-Quantum-Cryptography/EmailList
17. G.E. Sery, Approaching the one billion transistor logic product: Process and design challenges, in *Design, Process Integration, and Characterization for Microelectronics*. International Society for Optics and Photonics, vol. 4692, pp. 254–261, 2002
18. P.R. Groeneveld, Physical design challenges for billion transistor chips, in *Proceedings. IEEE International Conference on Computer Design: VLSI in Computers and Processors* (IEEE, 2002), pp. 78–83.
19. M.M. Waldrop, The chips are down for Moore's law. Nature News **530**(7589), 144 (2016)
20. Xilinx, *Vivado design suite user guide: High-level synthesis*, Accessed 8 Mar 2020, Dec. 2018. [Online]. Available: https://www.xilinx.com/support/documentation/sw_manuals/xilinx2018_3/ug902-vivado-high-level-synthesis.pdf
21. LegUp, *Legup user manual*, Accessed 8 Mar 2020, Oct. 2018. [Online]. Available: https://www.legupcomputing.com/docs/legup-6.2-docs/legup-6.2-docs.pdf
22. NEC, *CYBERWORKBENCH NEC's High Level Synthesis Solution*, Accessed 8 Mar 2020, Sep. 2016. [Online]. Available: https://www.nec.com/en/global/prod/cwb/pdf/CWB_Detailed_technical.pdf
23. Mentor Graphics, *Catapult C Synthesis User's and Reference Manual*, Accessed 8 Mar 2020, Nov 2010. [Online]. Available: http://readpudn.com/downloads377/ebook/1625374/catapult.useref.pdf
24. Cadence, *Stratus High-Level Synthesis*, Accessed 8 Mar 2020, Feb. 2015. [Online]. Available: http://www.europractice.stfc.ac.uk/vendors/cadence_stratus_ds.pdf

25. bluespec, *Bluespec SystemVerilog Reference Guide*, Accessed 8 Mar 2020, Jun. 2010. [Online]. Available: http://csg.csail.mit.edu/6.S078/6_S078_2012_www/resources/referenceguide.pdf
26. C. Lattner, V. Adve, Llvm: A compilation framework for lifelong program analysis & transformation, in *International Symposium on Code Generation and Optimization, 2004. CGO 2004* (IEEE, 2004), pp. 75–86
27. R. Kastner, J. Matai, S. Neuendorffer, Parallel Programming for FPGAs, ArXiv e-prints, May 2018. arXiv: 1805.03648
28. A. Mirhoseini, E.M. Songhori, F. Koushanfar, Idetic: A high-level synthesis approach for enabling long computations on transiently-powered asics, in *2013 IEEE International Conference on Pervasive Computing and Communications (PerCom)* (EEE, 2013), pp. 216–224
29. D. Soni, K. Basu, M. Nabeel, R. Karri, Power area, speed, and security (pass) trade-offs of nist pqc signature candidates using a c to asic design flow, in *IEEE International Conference on Computer Design (ICCD)*, 2019
30. J. Hennessy, D. Patterson, *Computer Architecture, A Quantitative Approach*, 5th edn. (Morgan Kaufmann Publishers, 2011). ISBN: 012383872X
31. ARM, *Power compiler user guide*, Accessed 16 Sep 2019, 2019. [Online]. Available: https://developer.arm.com/ip-products/physical-ip/embeddedmemory

Chapter 2
CRYSTALS-Dilithium

2.1 Algorithm Description

Dilithium is a lattice-based digital signature, built on the Learning with Errors (LWE) and Short Integer Solution (SIS) problems [1]. Dilithium is part of the Cryptographic Suite for Algebraic Lattices (CRYSTALS) submitted to the NIST PQC competition. Dilithium uses Fiat-Shamir with aborts [2] to improve upon the previous designs [3, 4], which use a uniform distribution and avoid Gaussian sampling, to achieve an optimized public key size—which is considered to be the main improvement. The scheme incorporates secure implementation against side-channel attacks and has comparable efficiency to the current best lattice-based signature schemes. Following previous work in [5], the security of Dilithium is proven in the Random Oracle Model (ROM), based on the hardness of two problems: (i) standard LWE and (ii) SelfTargetMSIS [6]. It is possible to get a non-tight reduction from SIS to SelfTargetMSIS using the forking lemma [7, 8] within ROM. Several works have provided evidence that Dilithium is secure in the QROM [9, 10]. Dilithium is Strong Unforgeability under Chosen Message Attacks (SUF-CMA) secure in the ROM, with a non-tight reduction. Dilithium has four variants spanning four security levels as outlined in Table 2.1, three of which correspond to NIST security levels 1, 2, and 3.

In the following sections, we discuss each of the components of Dilithium—key generation, signature generation, and signature verification.

Key Generation The key generation component produces the secret key for signature generation and the public key for signature verification (Algorithm 1). At the outset, AES creates seeds ρ, ρ', and a key. Using ρ, key generation expands a $(K \times L)$-size public matrix A. Either AES or SHAKE is chosen to implement the expansion of the matrix—depending upon the Dilithium variant. Each entry of matrix A is a polynomial in ring $R_q = \mathbb{Z}_q[X]/(X^n + 1)$ where $q = 2^{23} - 2^{13} + 1$ and $n = 256$. Similarly, ρ' and *nonce* act as the inputs for `shake256` to create

D. Soni et al., *Hardware Architectures for Post-Quantum Digital Signature Schemes*, https://doi.org/10.1007/978-3-030-57682-0_2

Table 2.1 Security parameters, signature size, public key size, and secret key size of Dilithium variants

	`Weak`	`Medium`	`Recommended`	`Very high`
NIST security level	–	1	2	3
Parameter q	8380417	8380417	8380417	8380417
Parameter d	14	14	14	14
Parameter weight of c	60	60	60	60
Parameter $\gamma_1 = (q-1)/16$	523776	523776	523776	523776
Parameter $\gamma_2 = \gamma_1/2$	261888	261888	261888	261888
Parameter (k, l)	(3,2)	(4,3)	(5,4)	(6,5)
Parameter η	7	6	5	3
Parameter β	375	325	275	175
Parameter ω	64	80	96	120
Signature size (bytes)	1387	2044	2701	3366
Public key size (bytes)	896	1184	1472	1760
Secret key size (bytes)	2081	2733	3348	3916

vectors s_1 and s_2 with a uniform distribution. The value of *nonce* is updated with each key of the various expansion functions for matrix A. The size of s_1 and s_2 are l and k, respectively. For matrix-vector multiplication of A and s_1, forward Number Theoretic Transform (NTT) is performed on s_1. The loop is run for a length of s_2's size (or k times). In each iteration of the loop, t stores the vector multiplication of one row of matrix A and s_1. `poly_reduce` reduces the coefficients of t. In the next step, inverse NTT is performed on t. When the loop is over, the matrix multiplication is completed and stored in t. Afterwards, t stores the sum of t and uniform distributed vector s_2. The coefficients of t will be reduced once again. t_0 and t_1 are separated from vector t, depending upon the coefficients. ρ and t_1 values are packed to create public key (pk). Function `shake256` produces an output tr with input pk. ρ, key, tr, s_1, s_2, and t_0 are combined to create secret key (sk). Key generation produces the secret key and public key.

Signature Generation The signature generation algorithm produces the signature using the message and secret key (Algorithm 2). First, seed values (ρ, key), tr, s_1, s_2, and t_0 are extracted from secret key sk. The input message m is copied in the last $mlen$ entries of signature (sm). The function `shake256` produces collision resistant hash μ, with tr and m as input. Following the same approach as key generation, matrix A is expanded using seed ρ. Forward NTT is calculated for s_1, s_2, and t_0. The remainder of the algorithm runs in an infinite while loop as follows.

Intermediate vector y is generated using part of the secret key and a harmonized nonce. After the intermediate polynomial is generated, the input nonce is incremented to ensure different intermediate values of y for each while loop. w stores the matrix-vector multiplication product of A and y. As shown in Algorithm 2, four functions are required to perform matrix multiplication: `polyvecl_ntt` performs forward NTT on y, `polyvecl_pointwise_acc_invmontgomery` multi-

Algorithm 1 CRYSTALS-Dilithium: Key generation

Input : No Input Required
Output : Secret Key $sk = (\rho, \text{key}, tr, s_1, s_2, t_0)$ and Public Key $pk = (\rho, t_1)$

procedure KEY GENERATION ▷ crypto_sign_keypair(sk, pk)
  $nonce \leftarrow 0$
  Generation of seed (ρ, ρ', key) using AES ▷ randombytes
  Expand matrix A from ρ. ▷ expand_mat
  Generation of short vectors s_1 and s_2 from ρ'. ▷ poly_uniform_eta
  $s_1 \leftarrow \text{NTT}(s_1)$. ▷ polyvecl_ntt
  for $(k = 0; k < lengthof s_2; k++)$ **do**
    $t \leftarrow A * s_1$ ▷ polyvecl_pointwise_acc_invmontgomery
    Reduce the coefficient of t. ▷ poly_reduce
    $t \leftarrow \text{NTT}^{-1}(t)$. ▷ poly_invntt_montgomery
  end for
  $t \leftarrow t + s_2$ ▷ polyveck_add
  Reduce the coefficient of t. ▷ polyveck_freeze
  $(t_0 + t_1) \leftarrow t$ ▷ polyveck_power2round
  $pk \leftarrow (\rho, t_1)$ ▷ pack_pk
  $tr \leftarrow$ `shake256`(pk) ▷ `shake256`
  $sk \leftarrow (\rho, \text{key}, tr, s_1, s_2, t_0)$ ▷ pack_sk
  return (sk, pk)
end procedure

plies two vectors, `poly_invntt_montgomery` performs inverse NTT on the result of vector multiplication, and `polyveck_freeze` reduces the coefficients of w. The resultant w is decomposed into two parts, w_1 and tmp, in such a way that $w_1 * \alpha + tmp = w$, for each element of the array (or coefficient of polynomial). Based on `shake256`, array c is generated with 60 nonzero elements with the values $[-1, 1]$. Following steps similar to those shown in signature generation, the corresponding algorithm performs $z = y + c * s_1$. If the infinity norm of the z is higher than $(\gamma_1 - \beta)$, the algorithm discards the signature and resumes at the beginning of the while loop. Similarly, $w' = w - c * s_2$ is calculated and w' is decomposed into $wcs20$ and tmp. If one of the reduced coefficients of $wcs20$ is greater than $\gamma_2 - \beta$, the algorithm discards the signature and continues with the next iteration of the while loop. If tmp is not equal to w_1, the while loop is restarted. If the above conditions are not true, $c * t_0$ is calculated. If the infinity norm of ct_0 is greater than $\gamma_2 - \beta$, the algorithm starts the while loop once again to generate a different signature. If infinity norm of ct_0 is less than $\gamma_2 - \beta$, $tmp = wcs_2 + ct_0$ is calculated and the tmp coefficients are reduced. The inputs tmp and ct_0 are provided to `polyveck_make_hint`, which produces the hint vector n. If n is not equal to ω, the signal is discarded. If hint vector n is equal to ω, the signature satisfies all the conditions. Hence, the (z, h, c) values are stored in sm and the length of the signature is updated. Finally, signature generation outputs signature (sm), which marks the end of signature generation and end of the while loop.

Algorithm 2 CRYSTALS-Dilithium: Signature generation

Input : Message m, Message Length $mlen$, and Secret Key $sk = (\rho, \text{key}, tr, s_1, s_2, t_0)$
Output : Signature $sm = (z, h, c)$, Signature Length $smlen$

procedure SIGNATURE GENERATION ▷ crypto_sign($sm, smlen, m, mlen, sk$)
  $nonce \leftarrow 0$
  Extract ρ, key, tr, s_1, s_2, t_0 from secret key ▷ unpack_sk
  Store the input message at the end of sm.
  $\mu \leftarrow CRH(tr, m)$ ▷ `shake256`
  Expand matrix A from ρ ▷ expand_mat
  $(s_1, s_2, t_0) \leftarrow \text{NTT}(s_1, s_2, t_0)$ ▷ polyvecl_ntt
  while 1 **do**
    Generate intermediate vector y ▷ poly_uniform_gamma1m1
    $w \leftarrow A * y$ ▷ polyvecl_ntt, polyvecl_pointwise_acc_invmontgomery, poly_invntt_montgomery, polyveck_freeze
    Decompose w, $(w_1, tmp) \leftarrow w$ ▷ polyveck_decompose
    $c \leftarrow \text{Hash}(\mu, w_1)$ ▷ challenge
    $z \leftarrow y + c * s_1$ ▷ poly_ntt, polyvecl_pointwise_invmontgomery, poly_invntt_montgomery, polyvecl_add, polyvecl_freeze
    if $(z > \gamma_1 - \beta)$ **then** Continue ▷ polyvecl_chknorm
    end if
    $w' \leftarrow w - cs_2$ ▷ polyvecl_pointwise_invmontgomery, poly_invntt_montgomery, polyvecl_sub, polyveck_freeze
    Decompose w', $(wcs20, tmp) \leftarrow w'$ ▷ polyveck_decompose
    Reduce the coefficient of $wcs20$. ▷ polyveck_freeze
    if $(wcs20 > \gamma_2 - \beta)||(tmp! = w_1)$ **then** Continue ▷ polyvecl_chknorm
    end if
    $ct_0 \leftarrow c * t_0$ ▷ polyvecl_pointwise_invmontgomery, poly_invntt_montgomery, polyveck_freeze
    if $(ct_0 > \gamma_2 - \beta)$ **then** Continue ▷ polyvecl_chknorm
    end if
    $tmp \leftarrow wcs_2 + ct_0$ ▷ polyveck_add
    Reduce the coefficient of tmp. ▷ polyveck_freeze
    $n \leftarrow (tmp, -ct_0)$ ▷ polyveck_neg, polyveck_make_hint
    if $(n! = \omega)$ **then** Continue
    end if
    $sm \leftarrow (z, h, c)$ ▷ pack_sig
    $smlen = mlen$ +CRYPTO_BYTES
    return (sm)
  end while
end procedure

Signature Verification The signature verification algorithm verifies whether the signature is from an authenticated user who has the corresponding, signing, secret key (Algorithm 3). The signature (z, h, c) and (ρ, t_1) are extracted from input signature (sm) and public key (pk), respectively. If the infinity norm of z is greater than $\gamma_1 - \beta$, the signature is rejected. Otherwise, the other condition for verification is checked using a hint vector. First, ρ, t_1, and m are provided as inputs to collision resistant hash (`shake256`), which produces μ. Similar to key generation and signature generation algorithms, the matrix A is expanded using seed ρ. Similarly, matrix multiplication and vector addition are performed, as mentioned in signature

Algorithm 3 CRYSTALS-Dilithium: Signature verification

Input : Signature $sm = (z, h, c)$, Message = m, Signature length = $smlen$ Public Key $pk = (\rho, t_1)$
Output : "Accept"/"Reject", Message m, Message Length $mlen$

procedure SIGNATURE VERIFICATION ▷ crypto_sign_open($m, mlen, sm, smlen, pk$)
  if smlen<CRYPTO_BYTES **then return** "Reject"
  end if
  $(\rho, t_1) \leftarrow pk$ ▷ unpack_pk
  $(z, h, c) \leftarrow sm$ ▷ unpack_sig
  if $z > (\gamma_1 - \beta)$ **then return** "Reject" ▷ polyvecl_chknorm
  end if
  $\mu \leftarrow CRH(CRH(\rho, t_1), m)$ ▷ shake_256, shake_256
  Expand matrix A. ▷ expand_mat
  $tmp1 \leftarrow A * z$ ▷ polyvecl_ntt, polyvecl_pointwise_acc_invmontgomery
  $tmp2 \leftarrow c * t_1$ ▷ poly_ntt, polyveck_shiftl, polyveck_ntt, poly_pointwise_invmontgomery
  $tmp1 \leftarrow tmp1 - tmp2$ ▷ polyveck_sub, polyveck_freeze, polyveck_invntt_montgomery
  $w_1 \leftarrow (temp1, h)$ ▷ polyveck_use_hint
  $cp \leftarrow H(\mu, w_1)$ ▷ challenge
  if $(cp! = c)$ **then return** "Reject"
  else
    $m \leftarrow sm$[CRYPTO_BYTES]
    return "Accept"
  end if
end procedure

generation, to calculate $tmp1 = A{*}z - c{*}t_1$. Using the `polyveck_use_hint` function, the algorithm calculates hint vector w_1. Based on `shake256` and inputs μ & w_1, the array cp is generated with 60 nonzero elements with the values $[-1, 1]$. If cp does not match with c (extracted from input signature), the signature is rejected. Otherwise, the last $mlen$ entries of signature (sm) are copied to message (m) and the signature is "Accepted."

Takeaway If the infinity norm of z is greater than $\gamma_1 - \beta$, the signature is rejected. If cp does not match with c (extracted from input signature), the signature is rejected. Only when the first condition is not satisfied AND the second condition produces a match, is the signature accepted.

2.2 Reference C Code ⟶ HLS-Ready C Code

CRYSTALS-Dilithium is the simplest and most flexible signature-based algorithm in terms of C design and implementation. It has the least changes required for key generation, signature generation, and signature verification. The changes to Dilithium's C code are common for all NIST PQC Round-2 digital signatures to

Table 2.2 CRYSTALS-Dilithium: changes to reference C code

Class	Number of changes
Remove a library function	5
Replace dynamic memory	3
Change file operation	3
Modify code flow	2
Total	13

make the C code capable of synthesis. The changes are divided into classes, as shown in Table 2.2.

- Dilithium uses AES to generate pseudo-random numbers; it uses "`memcpy`" and `AES256_ECB` function from library string.h and openssl, respectively. These functions are replaced with identical functionality.
- Key generation, signature generation, and signature verification have different inputs and outputs. These inputs/outputs, including the message, signature, public key, and secret key, have fixed memory size.
- File operations in the testbench also change accordingly.
- Signature verification has `goto` statements. We modify the code to avoid `goto` statements.

Takeaway Conversion of raw C code to HLS-ready C code requires modifying `goto` statements. The library functions in string.h and openssl library should also be replaced before HLS.

2.3 FPGA-Specific Implementations and Optimizations

The critical functions are identified for each of the three operations (keypair generation, signature generation, and signature verification). The time and area are distributed among many functions. Hence, we have performed optimizations for many functions/modules simultaneously (Figs. 2.1, 2.2, and 2.3).

There are many functions such as `ntt`, `nttinv`, functions in poly.c, and functions in polyvec.c, which are optimized. These optimizations on the baseline implementation provide the results shown in Tables 2.3, 2.4, and 2.5. These tables report the area overhead and latency for key generation, signature generation, and signature verification, respectively. The baseline version only optimizes the area overhead to fit the design in the Artix-7 board and it does not optimize performance (latency). Dilithium_very_high with security level 3 takes up maximum area for all three components. All the components for the different security levels fit into the Artix-7 board. The loop unrolling and pipelining optimizations improve the

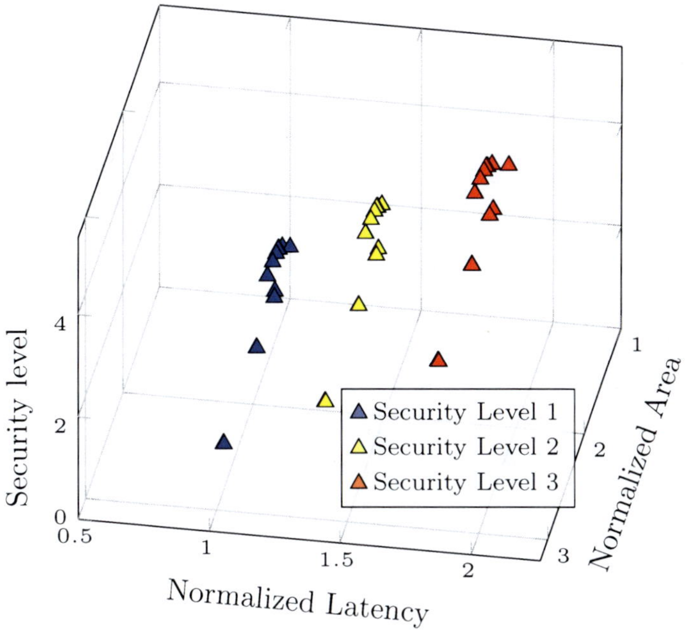

Fig. 2.1 FPGA design-space exploration of CRYSTALS-Dilithium key generation component using loop unrolling and loop pipelining optimizations. Latency and area are normalized with respect to security level 1 baseline implementation (which contains no performance optimizations)

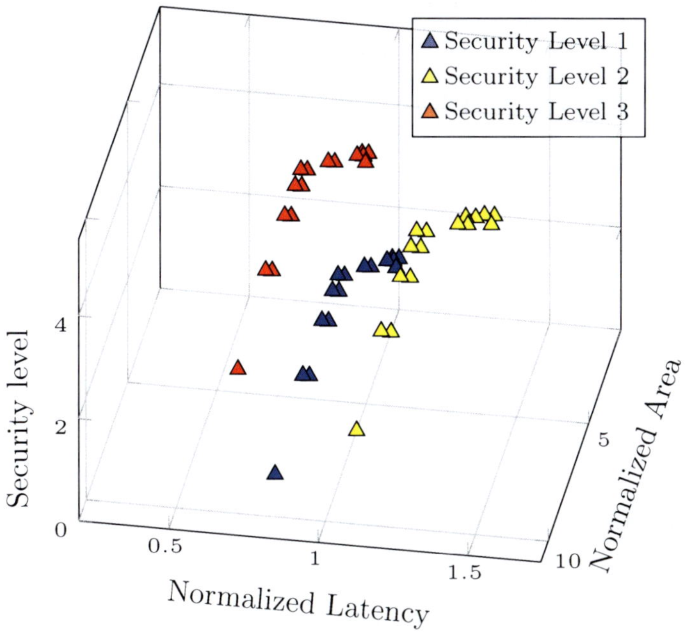

Fig. 2.2 FPGA design-space exploration of CRYSTALS-Dilithium signature generation component using loop unrolling and loop pipelining optimizations. Latency and area are normalized with respect to security level 1 baseline implementation (which contains no performance optimizations)

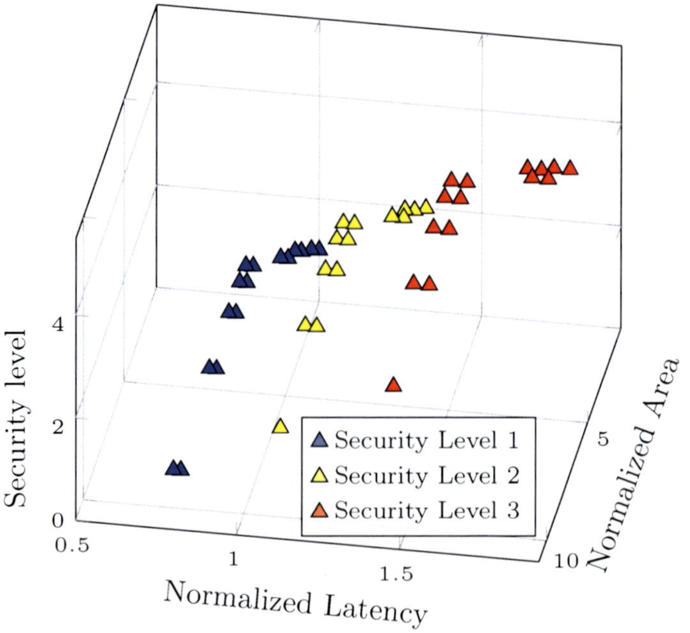

Fig. 2.3 FPGA design-space exploration of CRYSTALS-Dilithium signature verification component using loop unrolling and loop pipelining optimizations. Latency and area are normalized with respect to security level 1 baseline implementation (which contains no performance optimizations)

Table 2.3 Description: Performance, Area, and Security (PAS) trade-off of CRYSTALS-Dilithium key generation component for Artix-7 FPGA. Vivado-HLS directive loop unrolling and loop pipelining provide various designs, which have different latency and area requirements, for four CRYSTALS-Dilithium variants (`Dilithium-weak`, `Dilithium-medium`, `Dilithium-recommended`, and `Dilithium-very-high`) that have distinct security strength because of different algorithmic parameters

Variant	Sec. level	Optimization	FF	LUT	BRAM	DSP	Clock (ns)	Latency *Kcycles*
`Dilithium-weak`	–	Base	17783	86465	65	45	8.375	115
		Unroll	31401	127045	66	225	9.682	116
		Pipe	17310	86662	66	45	33.153	88
`Dilithium-medium`	1	Base	17627	86458	69	45	8.375	173
		Unroll	31491	127224	70	234	9.682	174
		Pipe	17634	86656	69	45	8.375	168
`Dilithium-recommended`	2	Base	17666	86448	80	45	8.375	241
		Unroll	31582	127196	81	225	9.682	241
		Pipe	17674	86646	80	45	8.375	233
`Dilithium-very-high`	3	Base	17864	87340	83	45	8.623	317
		Unroll	31843	128026	84	225	9.682	316
		Pipe	17872	87538	83	45	8.623	306

performance of the design by reducing the latency. Loop unrolling increases the performance at the cost of higher area overhead.

Table 2.4 Description: Performance, Area, and Security (PAS) trade-off of CRYSTALS-Dilithium signature generation component for Artix-7 FPGA. Vivado-HLS directive loop unrolling and loop pipelining provides various designs, which have different latency and area requirements, for four CRYSTALS-Dilithium variants (`Dilithium-weak`, `Dilithium-medium`, `Dilithium-recommended`, and `Dilithium-very-high`)

Variant	Sec. level	Optimization	FF	LUT	BRAM	DSP	Clock (ns)	Latency *Kcycles*
`Dilithium-weak`	–	Base	20912	89709	67	108	8.738	486
		Unroll	41295	129989	68	711	8.738	417
		Pipe	20980	90266	67	108	8.738	477
`Dilithium-medium`	1	Base	21023	89933	73	108	8.738	1260
		Unroll	41617	130632	74	720	8.738	1158
		Pipe	21094	90506	73	108	8.738	1232
`Dilithium-recommended`	2	Base	21089	89991	95	108	8.738	1660
		Unroll	42828	132103	96	711	8.738	1565
		Pipe	21160	90567	95	108	8.738	1618
`Dilithium-very-high`	3	Base	21265	91098	102	108	8.738	1133
		Unroll	43764	134037	103	711	8.738	1007
		Pipe	21322	91605	102	108	8.738	1106

Table 2.5 Description: Performance, Area, and Security (PAS) trade-off of CRYSTALS-Dilithium signature verification component for Artix-7 FPGA. Vivado-HLS directive loop unrolling and loop pipelining provide various designs, which have different latency and area requirements, for four CRYSTALS-Dilithium variants (`Dilithium-weak`, `Dilithium-medium`, `Dilithium-recommended`, and `Dilithium-very-high`)

Variant	Sec. level	Optimization	FF	LUT	BRAM (ns)	DSP *Kcycles*	Clock	Latency
`Dilithium-weak`	–	Base	15073	64930	42	72	8.738	147
		Unroll	43915	122802	44	702	9.83	118
		Pipe	15080	65147	42	72	8.738	144
`Dilithium-medium`	1	Base	15141	65074	47	72	8.738	215
		Unroll	44323	123447	48	711	9.83	176
		Pipe	15148	65293	47	72	8.738	210
`Dilithium-recommended`	2	Base	15161	65055	59	72	8.738	293
		Unroll	44329	123500	60	702	9.83	243
		Pipe	15169	65274	59	72	8.738	285
`Dilithium-very-high`	3	Base	15179	65141	61	72	8.738	381
		Unroll	44474	123746	62	702	9.83	317
		Pipe	15187	65360	61	72	8.738	370

As the security strength increases, area and latency for the algorithm should also increase. Hence, the highest security strength variant Dilithium_very_high should incur more latency and area overhead.

Takeaway Although Dilithium_very_high provides the strongest security, it incurs the highest latency and area overhead.

2.4 ASIC-Specific Implementations and Optimizations

In this section, the ASIC implementation results are discussed. Tables 2.6, 2.7, and 2.8 report the power, timing, and area for the various security levels of key generation, signature generation, and signature verification algorithms, respectively.

In Table 2.6, for all security levels, the latency of the loop-unrolled version of the algorithm is the same as the baseline, and the former presents a larger area and a higher clock period. Additionally, it should be noted that except for Dilithium_weak, the pipelined version has a very similar clock period and area as the baseline version and provides better latency numbers. Therefore, it makes sense to choose the pipelined version over the baseline and loop-unrolled versions thereof. Also in Table 2.6, dilithium_very_high presents the maximum area, whereas its standard cell area (Gate Count column) is almost the same as dilithium_medium, but the former has more RAM (in RAM size column), hence the total area.

In Table 2.7, although the latency for the loop-unrolled version is better than the baseline and the pipelined versions, its power consumption is almost double that of the other two. Moreover, the loop-unrolled version has larger area requirements. The pipelined and baseline versions have similar power, performance, and area requirements. Hence, it is advisable to choose the baseline version here as one can save tool execution time required to implement the pipelined version.

In Table 2.8, it is shown that the latency of the loop-unrolled version is better, and the associated clock period is comparable to the baseline and pipelined versions. The loop unrolled version could be chosen for the exhibited faster execution time ($Latency \times ClockPeriod$) but at the cost of high area requirements, since the loop unrolled version requires more than double the area, as compared to the baseline and pipelined versions.

Across the different scheme variants, dilithium_very_high has the highest total area required as it has more SRAM; for the given optimization technique, the standard cell area is comparable to other variants. As expected, the loop-unrolled version of each variant requires more area, hence higher power.

Table 2.9 reports high area utilizing internal modules for key generation, signature generation, and signature verification. Based on these reports, key design blocks in CRYSTALS-Dilithium are `shake_256`, `expand_mat`, `randombytes`, `ntt/nttinv/polyvecl_pointwise`, `poly_uniform_gamma1m`, and `challenge`.

Collision resistant hash function `shake_256`, which is a basic building block, is one of the most area-consuming modules for the scheme's three modules. Number Theory Transform-NTT (`ntt/invntt`) is another building block for

Table 2.6 Description: ASIC Power, Performance, and Area (PPA) of different security levels of Dilithium key generation algorithms. Gate Count is technology independent and is derived by computing $(TotalArea - MemoryArea) / (Nand2\ gate\ area)$. The Nand2 gate used to obtain the above results has an area of 2.1 μm^2 with GF 65 nm LPE process technology. Given a gate count and a memory size, the ASIC area can be estimated in any technology. FF Memory size in Kbytes gives the amount of memory implemented using Flip Flops instead of SRAM or SROM. Execution time can be derived by computing $(ClockPeriod \times Latency)$

Variant	Sec. level	Optimization	Power (mW)	Clock period (ns)	Area in μm^2 SRAM /ROM	Area in μm^2 Total area	Gate count $Kgate$	No of flip flops	Memory size in $Kbytes$ SRAM	Memory size in $Kbytes$ SROM	Memory size in $Kbytes$ FF	Latency $Kcycles$
`Dilithium-weak`	–	Base	3.68	4.27	475077	1179476	335.43	31740	20.45	3.94	1.57	115
		Unroll	4.47	4.66	492723	1291780	380.50	36711	20.45	3.94	1.57	116
		Pipe	5.84	9.59	497814	1119788	296.18	31229	20.45	3.94	1.57	88
`Dilithium-medium`	1	Base	3.75	4.29	638146	1348188	338.12	32288	40.01	3.94	1.62	173
		Unroll	4.62	4.65	655793	1470669	388.04	37376	40.01	3.94	1.62	174
		Pipe	3.74	4.39	638146	1348415	338.22	32280	40.01	3.94	1.62	168
`Dilithium-recommended`	2	Base	3.72	4.39	545810	1255396	337.90	32248	29.51	3.94	1.62	241
		Unroll	4.51	4.58	563456	1372899	385.45	37432	29.51	3.94	1.62	241
		Pipe	3.74	4.34	545810	1253575	337.03	32237	29.51	3.94	1.62	233
`Dilithium-very-high`	3	Base	3.80	4.39	727270	1443691	341.15	32494	51.95	3.94	1.62	317
		Unroll	4.64	4.60	744916	1559876	388.08	37598	51.95	3.94	1.62	316
		Pipe	3.81	4.20	727270	1438120	338.50	32486	51.95	3.94	1.62	306

Table 2.7 **Description:** ASIC Power, Performance, and Area (PPA) of different security levels of Dilithium signature generation algorithm. Gate Count is technology independent and is derived by computing $(TotalArea - MemoryArea) / (Nand2\ gate\ area)$. The Nand2 gate used to obtain the above results has an area of 2.1 μm^2 with GF 65 nm LPE process technology. Given a gate count and a memory size, the ASIC area can be estimated in any technology. FF Memory size in Kbytes gives the amount of memory implemented using Flip Flops instead of SRAM or SROM. Execution time can be derived by computing $(ClockPeriod \times Latency)$

Variant	Sec. level	Optimization	Power (mW)	Clock period (ns)	Area in μm^2 SRAM /ROM	Total area	Gate count $Kgate$	No of flip flops	Memory size in $Kbytes$ SRAM	SROM	FF	Latency $Kcycles$
`Dilithium-weak`	–	Base	4.92	4.56	699453.10	1488659.50	375.81	34035	41.12	3.59	1.46	486
		Unroll	9.04	5.10	729531.41	2008602.40	609.08	52953	41.31	4.31	1.27	417
		Pipe	4.90	4.39	699453.10	1485962.02	374.53	34059	41.12	3.59	1.46	477
`Dilithium-medium`	1	Base	5.07	4.51	985785.45	1776471.45	376.52	34220	74.99	3.59	1.46	1260
		Unroll	9.95	5.19	1017790.03	2343427.39	631.26	53940	75.19	4.31	1.27	1158
		Pipe	5.01	4.31	985785.45	1776090.21	376.34	34246	74.99	3.59	1.46	1232
`Dilithium-recommended`	2	Base	5.00	4.35	825784.40	1616279.24	376.43	34144	57.34	3.59	1.46	1660
		Unroll	10.20	5.15	856853.67	2180653.95	630.38	53292	57.53	4.31	1.27	1565
		Pipe	4.97	4.87	825784.40	1619415.56	377.92	34160	57.34	3.59	1.46	1618
`Dilithium-very-high`	3	Base	5.15	4.40	1136803.49	1931073.28	378.22	34363	94.09	3.59	1.46	1133
		Unroll	9.33	4.67	1169743.38	2505539.58	636.09	54518	94.28	4.31	1.27	1007
		Pipe	5.14	4.60	1136803.49	1930528.24	377.96	34389	94.09	3.59	1.46	1106

Table 2.8 **Description:** ASIC Power, Performance, and Area (PPA) of different security levels of the Dilithium signature verification algorithm. Gate Count is technology independent and is derived by computing $(TotalArea - MemoryArea)$ / $(Nand2\ gate\ area)$. The Nand2 gate used to obtain the above results has an area of 2.1 μm^2 with GF (Gate Forest Technology) 65 nm Low-Power Enhanced (LPE) process technology. Given a gate count and a memory size, the ASIC area can be estimated in any technology. FF Memory size in Kbytes gives the amount of memory implemented using Flip Flops instead of SRAM or SROM. Execution time can be derived by computing $(ClockPeriod \times Latency)$

Variant	Sec. level	Optimization	Power (mW)	Clock period (ns)	Area in μm^2 SRAM /ROM	Total area	Gate count $Kgate$	No of flip flops	Memory size in $Kbytes$ SRAM	SROM	FF	Latency $Kcycles$
Dilithium-weak	–	Base	2.54	4.29	337897.29	745329.09	194.02	17037	19.90	2.88	0.59	147
		Unroll	5.89	4.70	431500.62	1576296.30	545.14	49846	20.36	2.88	0.65	118
		Pipe	2.56	4.14	337897.29	743911.05	193.34	17028	19.90	2.88	0.59	144
Dilithium-medium	1	Base	2.64	4.87	476767.53	885565.52	194.67	17132	38.40	2.88	0.59	215
		Unroll	6.01	4.66	585993.57	1742365.41	550.65	50261	38.86	2.88	0.65	176
		Pipe	2.64	4.18	476767.53	885388.04	194.58	17124	38.40	2.88	0.59	210
Dilithium-recommended	2	Base	2.60	4.54	397189.01	805758.77	194.56	17103	28.43	2.88	0.59	293
		Unroll	5.99	4.66	497924.49	1645309.29	546.37	50233	28.89	2.88	0.65	243
		Pipe	2.63	4.02	397189.01	805992.05	194.67	17092	28.43	2.88	0.59	285
Dilithium-very-high	3	Base	2.66	4.45	551070.56	961782.68	195.58	17152	49.80	2.88	0.59	381
		Unroll	6.08	4.65	669741.91	1833552.07	554.20	50411	50.27	2.88	0.65	317
		Pipe	2.68	4.22	551070.56	960654.44	195.04	17144	49.80	2.88	0.59	370

Table 2.9 Description: Area consumed by the design blocks used in Dilithium

	% of total area											
	Dilithium-weak			Dilithium-medium			Dilithium-recommended			Dilithium-very-high		
Design blocks	Base	Unroll	Pipe	Base	Unroll	Pipe	Base	Unroll	Pipe	Base	Unroll	Pipe
Key generation												
shake_256	28.0	27.9	25.9	28.6	26.1	28.5	26.6	24.3	26.6	24.9	22.9	24.7
expand_mat	14.6	13.3	13.4	13.7	12.5	13.8	12.8	11.7	12.8	12.0	11.0	12.0
randombytes	9.7	8.9	10.2	9.4	8.6	9.5	8.8	8.1	8.8	8.2	7.6	8.2
ntti/nttinv/polyvecl_pointwise	7.4	13.7	8.8	7.3	13.7	7.1	6.7	13.1	6.6	6.3	12.4	6.2
Signature generation												
poly_uniform_gamma1m	11.8	8.5	11.8	10.9	7.8	10.9	9.9	7.3	9.9	9.1	7.1	7.1
expand_mat	11.4	8.1	11.4	10.5	7.5	10.5	9.5	7.0	9.5	8.8	6.8	6.8
challenge	11.4	8.6	11.4	10.6	8.0	10.6	9.6	7.5	9.7	9.0	7.4	7.4
shake_256	10.9	7.6	10.9	10.0	6.9	10.1	9.1	6.5	9.1	8.4	29.3	10.1
ntti/nttinv/polyvecl_pointwise	13	36.6	13	12.0	35.7	11.9	11.1	32.5	11	10.1	6.4	8.4
Signature verification												
expand_mat	22.7	10.7	22.7	21	10.2	21.0	19.1	9.7	19.1	17.7	9.2	17.7
shake_256	21.4	19.7	21.4	19.8	18.8	19.9	18.1	17.8	18.1	16.7	17	16.7
ntti/nttinv/polyvecl_pointwise	21	48.3	20.9	19.3	46.5	19.4	17.8	44.3	17.6	16.4	42.4	16.3

Dilithium, which occupies a large area. Dilithium utilizes NTT to perform complex computation. `Expand_mat`, used by three modules, is a high area-consuming module because it expands the common matrix A. The pseudo-random number generator (`randombytes`), essential for generating a pseudo-random looking key, has the third-largest area overhead for key generation. The challenge function, shown in Table 2.9, uses `shake_256` implementation to generate a stream of hash output. As this function calls `shake_absorb` and `shake_squeezeblock` separately for `shake_256` implementation, the tool generates a separate hardware design (Figs. 2.4, 2.5, and 2.6).

Takeaway CRYSTALS-Dilithium at security level 3 is a better candidate for implementation as it provides better security with minimal overhead in latency and area.

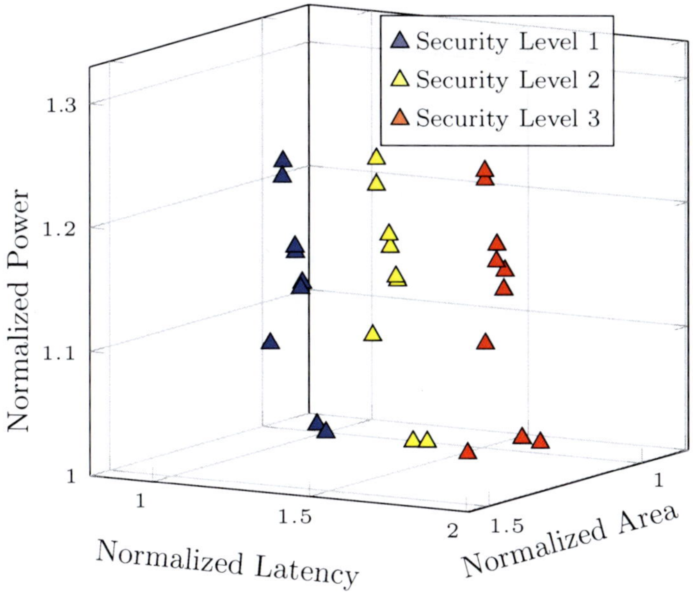

Fig. 2.4 ASIC design-space exploration of CRYSTALS-Dilithium key generation component when implemented using loop unrolling and pipelined optimizations, normalized to a baseline of security level 1 key generation

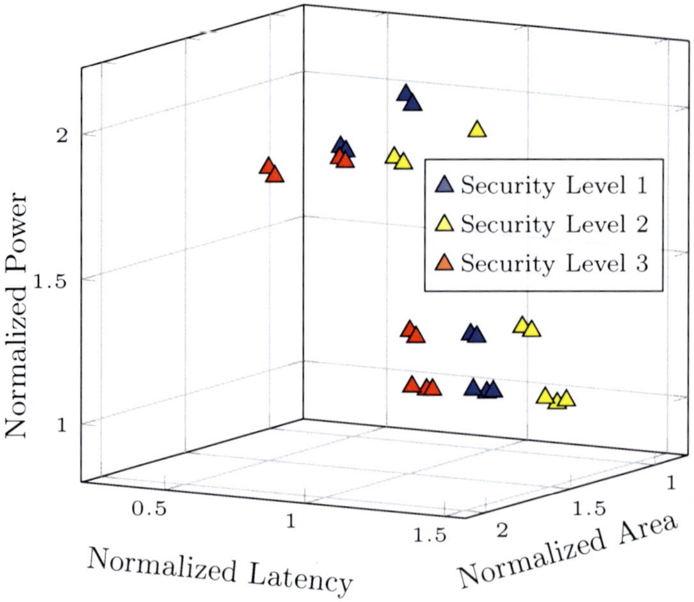

Fig. 2.5 ASIC design-space exploration of CRYSTALS-Dilithium signature generation component when implemented using loop unrolling and pipelined optimizations, normalized to a baseline of security level 1 signature generation

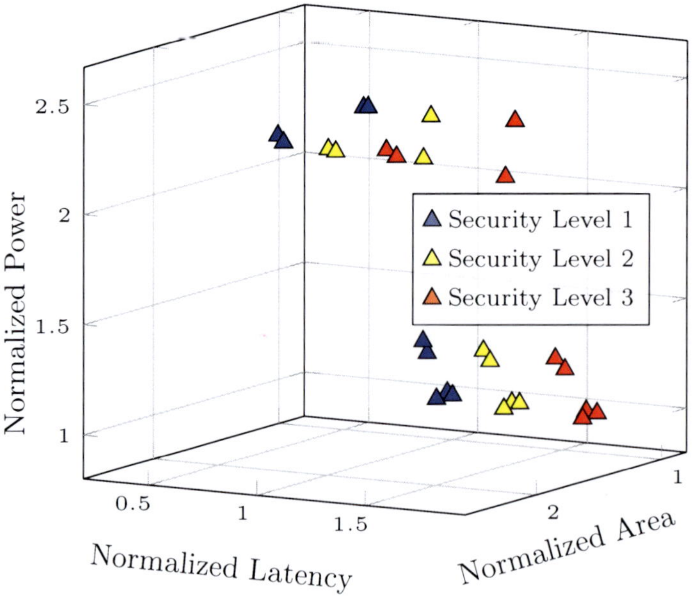

Fig. 2.6 ASIC design-space exploration of CRYSTALS-Dilithium signature verification component when implemented using loop unrolling and pipelined optimizations, normalized to a baseline of security level 1 signature verification

2.5 Key Takeaways

- Clock period for all the components in FPGA are around 9.83 ns. Hence the frequency of CRYSTALS-Dilithium is more than 100 MHz.
- As the security strength increases, the length of the secret and public keys increase; the latency for Dilithium increases with the security strength. However, the area overhead is similar for all security levels.
- Loop pipelining and loop unrolling are the main optimizations performed for design-space exploration. Loop unrolling increases the clock period. Hence, the baseline implementation is a better implementation for key generation.
- Signature generation at security level 3 improves the performance, using more area overhead compared to other security levels.
- Signature generation at security level 3 has better performance than security levels 1 and 2. Due to higher security and better performance, security level 3 is a better candidate for implementation.
- For signature verification, loop unrolling provides better performance for all security levels. This is visible through security level 3 design points. The latency improves significantly as area increases.
- Loop unrolling and loop pipelining offer no improvements for key generation. However, they provide efficient implementation for signature verification.
- In ASIC implementation, the average clock period across the variants is 4.73 ns, which corresponds to 211 MHz. When compared to 28 nm FPGA frequency numbers in Table 2.3, ASIC in 65 nm technology is about 85% faster for the Dilithium designs.
- The number of flip flops is more in ASIC than the FPGA equivalent. This is because all the memory is implemented using BRAM in FPGA. But in ASIC, the memory is converted to SRAM only if the conversion gives area gains; thus, smaller memory is not implemented as SRAM, but with flip flops.
- For ASIC, the security level 3 variant is a better candidate for implementation as it provides better security with minimal overhead in latency and area.
- Power overhead does not always increase with the security level. Furthermore, the design optimizations that parallelize the implementations also increase the power consumption.

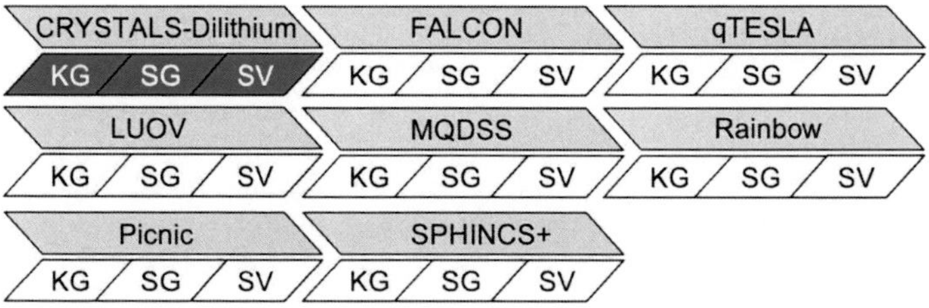

References

1. V. Lyubashevsky, L. Ducas, E. Kiltz, T. Lepoint, P. Schwabe, G. Seiler, D. Stehle, *Crystals-Dilithium*. Submission to the NIST Post-Quantum Cryptography Standardization Project, 2019. https://csrcnistgov/CSRC/media/Projects/Post-Quantum-Cryptography/documents/round2/submissions/CRYSTALSDilithiumRound2.zip
2. V. Lyubashevsky, Fiat-shamir with aborts: Applications to lattice and factoring-based signatures, in *International Conference on the Theory and Application of Cryptology and Information Security*, pp. 598–616, Dec. 2009
3. T. Güneysu, V. Lyubashevsky, T. Pöppelmann, Practical lattice-based cryptography: A signature scheme for embedded systems, vol. 7428, pp. 530–547, Sep. 2012. https://doi.org/10.1007/9783642330278_31
4. S. Bai, S. Galbraith, An improved compression technique for signatures based on learning with errors, Feb. 2014. https://doi.org/10.1007/978-3-319-04852-9_2
5. V. Lyubashevsky, Lattice signatures without trapdoors, in *Annual International Conference on the Theory and Applications of Cryptographic Techniques* (Springer, 2012), pp. 738–755
6. E. Kiltz, V. Lyubashevsky, C. Schaffner, A concrete treatment of fiat- shamir signatures in the quantum random-oracle model, in *Annual International Conference on the Theory and Applications of Cryptographic Techniques* (Springer, 2018), pp. 552–586
7. D. Pointcheval, J. Stern, Security arguments for digital signatures and blind signatures. J. Cryptol. **13**(3), 361–396 (2000)
8. M. Bellare, G. Neven, Multi-signatures in the plain public-key model and a general forking lemma, in *Proceedings of the 13th ACM Conference on Computer and Communications Security*, pp. 390–399, 2006
9. Q. Liu, M. Zhandry, Revisiting post-quantum fiat-shamir, in *Annual International Cryptology Conference* (Springer, 2019), pp. 326–355
10. J. Don, S. Fehr, C. Majenz, The measure-and-reprogram technique 2.0: Multi-round fiat-shamir and more. Preprint (2020). arXiv:2003.05207

Chapter 3
FALCON

3.1 Algorithm Description

FALCON, which stands for *Fast Fourier Lattice-based Compact signatures Over NTRU*, is a lattice-based cryptographic digital signature algorithm based on NTRU[1]. FALCON is designed with the intention of providing small public key and signature sizes in order to better facilitate the transition to post-quantum schemes. FALCON is an instantiation of the GPV framework [2] for constructing a hash-and-sign lattice-based signature scheme, for which the authors chose NTRU lattices and devised a practical instantiation of the fast Fourier trapdoor sampler from the work by Ducas and Prest [3]. FALCON is SUF-CMA-secure in both the ROM and QROM [2, 4], and its underlying hard problem is the Short Integer Solution problem (SIS) over NTRU lattices. FALCON has three variants, which are described in Table 3.1.

FALCON has nondeterministic key and signature generation modules. On the other hand, signature verification is deterministic. Now, we will examine the three modules in detail.

Key Generation Algorithm 1 shows FALCON key pair generation, which requires the computation of polynomials f, g, F, G, and h, as per the following equations:

$$fG - gF = q \; mod \; \phi, \tag{3.1}$$

$$h = gf^{-1} \; mod \; (\phi, q). \tag{3.2}$$

In order to calculate these, a pseudo-random number is generated by AES, which acts as a seed to initialize `SHAKE-256`. Using SHAKE-based pseudo-random numbers, the algorithm generates random polynomials f and g with a Gaussian distribution. If the squared norm of these polynomials is out of bound, they are discarded and new polynomials are generated. Similarly, the algorithm discards the

D. Soni et al., *Hardware Architectures for Post-Quantum Digital Signature Schemes*, https://doi.org/10.1007/978-3-030-57682-0_3

Algorithm 1 FALCON: Key generation

Input : No Input Required.
Output : Secret Key $sk = (f, g, F, G)$ and Public Key $pk = (h)$.

procedure KEY GENERATION ▷ crypto_sign_keypair(sk, pk)
    Generate seed for hash function using AES. ▷ randombytes
    Initialize the hash function using the seed. ▷ falcon_keygen_set_seed
    while (1) **do**
        Generate polynomial f and g. ▷ poly_small_mkgauss
        Check the norm of the polynomial. ▷ poly_small_sqnorm
        Compute the orthogonalized vector norm. ▷ "Multiple functions"
        if (vector norm < 16822) **then** ▷ fpt_lt
            "Continue"
        end if
        if (Fail to generate h) **then** ▷ falcon_compute_public
            "Continue"
        end if
        if (Fail to generate F and G) **then** ▷ solve_NTRU
            "Continue"
        end if
        "Break"
    end while
    Encode secret key. ▷ falcon_encode_small
    Encode public key. ▷ falcon_encode_12289
    return ($sk = (f, g, F, G)$, $pk = (h)$)
end procedure

polynomials if the orthogonalized vector norms of the polynomials are out of bound. To compute the orthogonalized vector norm, the algorithm uses the Fast-Fourier Transform (FFT). Using f and g polynomials, the algorithm computes public key polynomial h such that it satisfies Eq. 3.2. The key generation component solves the Eq. 3.1 (NTRU equation) to compute polynomials F and G. For the secret key, the algorithm encodes f, g, F, and G polynomials sequentially. The algorithm encodes the public key by encoding polynomial h. Finally, this algorithm produces the secret and public keys.

Table 3.1 Security parameters, signature size, and public key size of FALCON variants

	FALCON-512	FALCON-768	FALCON-1024
NIST security level	1	2-3	4-5
Parameter n	512	768	1024
Parameter ϕ	$x^n + 1$	$x^n - x^{n/2} + 1$	$x^n + 1$
Parameter q	12289	18433	12289
Parameter β^2	43533782	100464491	87067565
Signature size (bytes)	657.38	993.91	1273.31
Public key size (bytes)	897	1441	1793

Algorithm 2 FALCON: Signature generation

Input : Message m, Message Length $mlen$, Secret Key $sk = (f, g, F, G)$.
Output : Signature sm = (sig_len, nonce, message, signature), Signature Length $smlen$

procedure SIGNATURE GENERATION ▷ crypto_sign($sm, smlen, m, mlen, sk$)
  Generate seed for hash function using AES. ▷ randombytes
  Initialize the hash function using the seed. ▷ falcon_sign_set_seed
  Decode and pre-computation of secret key. ▷ falcon_sign_set_private_key
  $c \leftarrow H(r, m)$. ▷ falcon_sign_start, falcon_sign_update
  $s_1 + s_2 h = c \ mod \ q$.
  Encode s_2. ▷ falcon_sign_generate
  Copy nonce and encoded s_2 to sm. ▷ memcpy
  return sm = (sig_len, nonce, message, encoded s_2).
end procedure

Signature Generation Algorithm 2 shows the signature generation steps for FALCON. First, a pseudo-random seed is generated for the hash function using AES. Using this seed, the algorithm initializes `SHAKE-256`. The signature generation algorithm computes a hash digest $c \in Z_q[x]/(\phi)$ from the salt r and input message m. The algorithm decodes the secret key (previously encoded in the key generation algorithm) and retrieves polynomials f, g, F, and G. If G is not extracted from the secret key, G is calculated. Then, it uses polynomials f, g, F, and G to calculate two short vectors s_1 and s_2 such that $s_1 + s_2 h = c \ mod \ q$, without leaking the secret key. It encodes short vector s_2 and then concatenates the signature length, salt, message, and encoded s_2. This concatenation of values is stored in sm.

Algorithm 3 FALCON: Signature verification

Input : Signature sm = (sig_len, nonce, message, encoded s_2), Signature length = $smlen$, Public Key $pk = (h)$.
Output : "Accept"/"Reject", Message m, Message Length $mlen$.

procedure SIGNATURE VERIFICATION ▷ crypto_sign_open($m, mlen, sm, smlen, pk$)
  Decode public key. ▷ falcon_vrfy_set_public_key
  if ($smlen <$ (2+PARAM_NONCE) || $sig\_len > (smlen -$ (2+PARAM_NONCE))) **then**
    return "Reject"
  end if
  Initialize the hash function. ▷ falcon_vrfy_start
  $c \leftarrow H(r, m)$ ▷ falcon_vrfy_update
  $s_2 \leftarrow$ Decode(signature). ▷ falcon_decode_small
  $s_1 \leftarrow c - s_2 h \ mod \ q$ ▷ falcon_vrfy_verify_raw
  if ($\|s_1, s_2\| > \beta$) **then** ▷ falcon_is_short
    return "Reject"
  end if
  return {"Accept", m, $mlen$}
end procedure

Signature Verification Algorithm 3 shows the signature verification steps for FALCON. The signature verification algorithm initially decodes the public key and extracts polynomial h. If the signature length is less than 2+PARAM_NONCE or sig_len is larger than $smlen$ − (2+PARAM_NONCE)), the signature is not valid. If the signature is valid, the algorithm initializes the hash function and computes a hash value $c \in Z_q[x](\phi)$. Then, it decodes the signature and extracts a short vector s_2; using s_2 and polynomial h, the algorithm calculates vector s_1. If L2-norm of aggregate vector (s_1, s_2) is greater than the bound, the signature is rejected; otherwise it is verified.

Takeaway FALCON is the only NIST PQC Round-2 signature scheme which requires floating-point operations. It also uses recursive functions.

3.2 Reference C Code ⟶ HLS-Ready C Code

FALCON uses floating point operations and recursive functions. Hence, we could not generate hardware for key generation and signature generation; we have only converted the signature verification component to HLS-ready C code. While we were verifying signature verification using C/RTL co-simulation, Vivado HLS threw GCC compilation errors because of recursive operations in FALCON. As a result, we had to remove all the recursive functionality to verify the signature verification hardware design. Table 3.2 classifies all the changes for FALCON signature verification into six categories, which we describe here.

- FALCON signature verification uses `memcpy` and `memset` functions in `shake.c`. We replace these library functions with identical functionality.
- The inputs and outputs of the three components are the message, signature, public key, and secret key. We express these inputs and outputs as fixed size arrays.
- We remove the "*restrict*" keyword used to provide hints to the compiler about pointer optimizations.
- We change the file operations of the testbench.
- Listings 3.1 and 3.2 show code modifications for the multiplication operation.
- We remove all the ternary operation code for `FALCON-512` and `FALCON-1024`. We also remove all the binary operation code for `FALCON-768`.

Table 3.2 FALCON: Changes to the reference C code

Class	Number of changes
Replace a library function	9
Remove dynamic memory allocation	3
Modify complex pointer operation	4
Change file operation	3
Modify code structure	2
Optimizations	4
Total	25

Takeaway Recursive operations in the C-implementation of FALCON had to be resolved to make it capable of synthesis before applying HLS.

3.3 FPGA-Specific Implementations and Optimizations

We have resolved a few errors in order to convert the reference C code into HLS-ready C code for the three variants of FALCON. Listings 3.1 and 3.2 show an important example of code modification for HLS. They describe the Montgomery multiplication operation code snippet from FALCON. Listing 3.1 details that *mq_montymul* requires multiplication, addition, AND operation, and shift operation. While calculating Listing 3.1-Line 4, the HLS tool does not consider 16 LSB of variable z and w. Besides this, the same function is used to produce 16-bit and 32-bit unsigned integers. Thus, the RTL generated by HLS is incorrect. To rectify this error we added a separate operation for the lower 16 bits of the inputs. We also created two separate functions to deliver an output of different sizes. These rectifications are detailed in Listing 3.2.

We optimize `NTT`, `inverse-NTT`, and `SHAKE` functions. While `NTT` and `inverse-NTT` functions are optimized using loop unrolling and loop pipelining, the `SHAKE` hash function is optimized using *Inlining* and *Allocation* directives.

Table 3.3 reports the results of the optimizations presented above. While `FALCON-512` and `FALCON-1024` use binary operations, `FALCON-768` uses ternary operations. As a result, the latter has different area and timing requirements compared to the former two. For the baseline implementation, `FALCON-768` requires the least number of flip-flops and LUTs among all three variants. However, it incurs the highest overhead in terms of DSPs and BRAMs. Loop pipelining improves latency by increasing the clock period. Loop unrolling slightly improves the performance with an additional $\sim$20% area overhead.

```
static inline uint32_t
mq_montymul(uint32_t x, uint32_t y, uint32_t q, uint32_t q0i)
{
        uint32_t z, w;
        z = x * y;
        w = ((z * q0i) & 0xFFFF) * q;
        z = (z + w) >> 16;
        z -= q;
        z += q & -(z >> 31);
        return z;
}
```

Listing 3.1 Mongtomery multiplication

```
static inline uint32_t
mq_montymul(uint32_t x, uint32_t y, uint32_t q, uint32_t q0i)
{
        uint32_t z, w, temp;
       z = (x * y + ((x * y * q0i)& 0xFFFF)*q) >> 16;
       if(z > q)
               z -= q;
       return z;
}
static inline uint16_t
mq_montymul16(uint32_t x, uint32_t y, uint32_t q, uint32_t q0i)
{
        uint32_t z, w;
       z=((x * y) >> 16) + ((((x * y * q0i)& 0xFFFF)*q) >> 16) +
    ((((x * y) & 0xFFFF) + ((((x * y * q0i)& 0xFFFF)*q) & 0xFFFF)
    )>>16);
       if(z > q)
               z -= q;
       return z;
}
```

Listing 3.2 Synthesizable Montgomery multiplication

Figure 3.1 shows the design-space exploration for different architectures of FALCON signature verification. Latency and area of the graph are normalized with security level 1 baseline implementation (i.e., `FALCON-512`). Figure 3.1 reveals improvements in area and latency for various designs because of the optimization directives. Security level 3 (i.e., `FALCON-768`) has the lowest area overhead while security level 1 has the lowest latency.

Table 3.3 Description: Performance, area, and security (PAS) trade-off of the FALCON signature verification component for Artix-7 FPGA. Vivado-HLS directives for loop unrolling and loop pipelining give different designs, which have different latency and area requirements, for three FALCON variants (`FALCON-512`, `FALCON-768`, and `FALCON-1024`)

Variant	Sec. level	Optimization	FF	LUT	BRAM	DSP	Clock (ns)	Latency *Kcycles*
`FALCON-512`	1	Base	17764	57604	18	26	12.939	77
		Unroll	21844	72325	18	89	12.939	77
		Pipe	17446	57805	18	26	32.434	62
`FALCON-768`	3	Base	16968	49151	19	79	12.928	124
		Unroll	19075	60065	19	43	13.018	132
		Pipe	16640	49330	19	79	35.292	105
`FALCON-1024`	5	Base	18241	58645	18	28	13.064	161
		Unroll	22224	73385	18	91	13.064	160
		Pipe	17865	58822	18	28	35.326	127

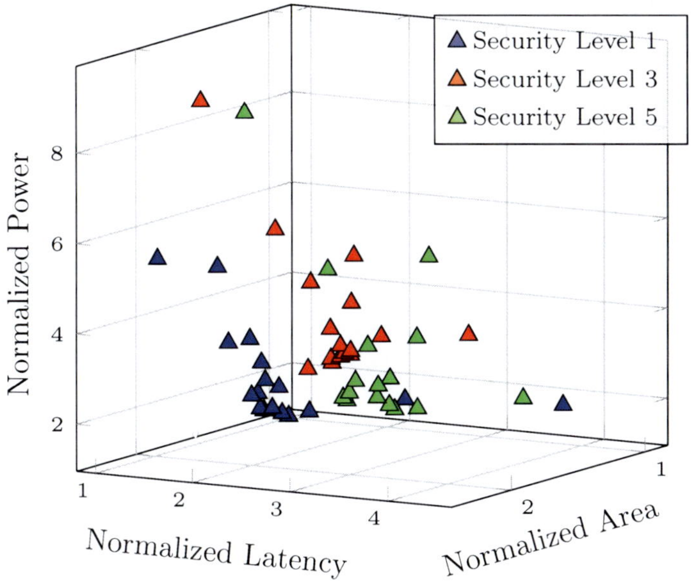

Fig. 3.1 FPGA design-space exploration of the FALCON signature verification component using loop unrolling and loop pipelining optimizations. Latency and area are normalized with respect to security level 1 baseline implementation (which contains no performance optimizations)

Takeaway The latency for the FALCON signature verification increases with the security strength.

3.4 ASIC-Specific Implementations and Optimizations

In Table 3.4, for all the three variants, baseline implementation provides better throughput ($ClockPeriod \times Latency$). The baseline version has better area and power for FALCON-512 and FALCON-1024. However, for FALCON-768, the pipelined version has slightly better area and power.

Table 3.5 reports the area utilization by the internal FALCON modules for signature verification. Based on these observed areas, key design blocks in FALCON are *vrfy_verify_r*, *hash_to_point*, *vrfy_update*, *vrfy_start*, and *mq_NTT_ternary_1*.

vrfy_verify_r takes the signature and hash values of a randomized input and the message to output validity of the signature. *hash_to_point* produces a new point using *shake_extract*. *vrfy_update* updates the input and internal states of SHAKE. *vrfy_start* initializes SHAKE, and finally, *mq_NTT_ternary_1* computes the NTT on a ring for the FALCON-768 variant.

Figure 3.2 shows the ASIC design-space exploration for different architectures of FALCON signature verification. Latency and area of the graph are normalized with security level 1 baseline implementation (i.e., FALCON-512).

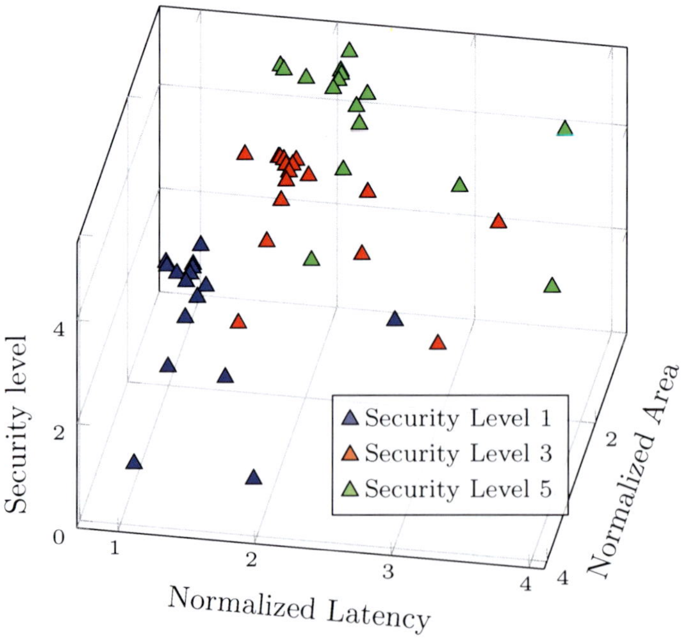

Fig. 3.2 ASIC design-space exploration of FALCON signature verification component. Latency and area are normalized with respect to security level 1 baseline implementation

Table 3.4 Description: ASIC Power, Performance, and Area (PPA) of different variants of the FALCON signature verification algorithm. Gate Count is technology independent and is derived by computing $(TotalArea - MemoryArea)$ / $(NAND2\ gate\ area)$. The NAND2 gate used to obtain the above results has an area of 2.1 μm^2 with GF (Gate Forest Technology) 65 nm Low-Power Enhanced (LPE) process technology. Given a gate count and a memory size, the ASIC area can be estimated in any technology. FF Memory size in kilobytes gives the amount of memory implemented using Flip Flops instead of SRAM or SROM. Execution time can be derived by computing $(ClockPeriod \times Latency)$

Variant	Security level	Optimization	Power (mW)	Clock period (ns)	Area in μm^2 SRAM /ROM	Total area	Gate count $Kgate$	No of flip flops	Memory size in $Kbytes$ SRAM	SROM	FF	Latency $Kcycles$
FALCON-512	1	Base	2.18	5.99	147757	530860	182.43	19388	8.07	5.25	0.20	77
		Unroll	3.65	6.38	147757	672540	249.90	22756	8.07	5.25	0.20	77
		Pipe	2.44	8.15	166903	553930	184.30	19156	8.07	5.25	0.20	62
FALCON-768	3	Base	5.43	5.17	198980	590957	186.66	18584	8.07	4.22	0.20	124
		Unroll	6.13	5.16	198980	597419	189.73	20453	8.07	4.22	0.20	132
		Pipe	4.81	6.32	198980	587742	185.13	18225	8.07	4.22	0.20	105
FALCON-1024	5	Base	2.97	4.24	147757	528751	181.43	19941	8.07	5.25	0.20	161
		Unroll	4.86	5.09	147757	677224	252.13	23600	8.07	5.25	0.20	160
		Pipe	3.27	5.79	166903	547253	181.12	19658	8.07	5.25	0.20	127

Table 3.5 Description: Area consumed by the design blocks used in the FALCON

	% of total area								
	FALCON-512			FALCON-768			FALCON-1024		
Design blocks	Base	Unroll	Pipe	Base	Unroll	Pipe	Base	Unroll	Pipe
Signature verification									
vrfy_verify_r	21.3	29.8	24.7	30.5	30.7	30.8	21.5	30.6	20.7
hash_to_point	17.1	13.7	20.4	15.4	15.4	15.4	17.2	13.7	16.5
vrfy_update	16.2	13.3	14.6	14.3	14.7	14.3	16.0	13.1	15.4
vrfy_start	16.2	13.3	10.5	NA	NA	NA	16.0	13.1	15.4
mq_NTT_ternary_1	NA	NA	NA	11.2	10.2	10.7	NA	NA	NA

3.5 Key Takeaways

- Minimum clock period for FALCON components is 32.434 ns. FALCON could run at a maximum frequency of ~30 MHz.
- For signature verification, the FALCON-1024 area overhead is similar to that for FALCON-512.
- FALCON-1024 latency is 3× higher than FALCON-512, thus FALCON-512 is better suited for faster implementations.
- As shown in Table 3.3, loop unrolling does not improve latency.
- As reported in Table 3.3, loop pipelining improves the latency by ~22% without additional memory or area cost. A loop-pipelined design has a clock period of 32.43 ns which is over 2× the clock period for the baseline and loop unrolled designs.
- Figure 3.1 shows the latency difference in loop pipelining and loop unrolling optimizations for FALCON-1024.
- The number of flip flops is more in ASIC than the FPGA equivalent. This is because all the memory is implemented using BRAM in FPGA. But in ASIC, the memory is converted to SRAM only if the conversion gives area gains; thus, smaller memory is not implemented as SRAM, but with flip flops.
- Table 3.3 shows that for pipelined architecture, the clock frequency for the ASIC implementation in 65 nm technology is 120% faster for FALCON-512, 150% faster for FALCON-768, and 200% faster for FALCON-1024 when compared to the corresponding 28nm FPGA implementations.
- As of April 2020, the supported FALCON versions are FALCON-512 and FALCON-1024.
- In this chapter, we evaluated and explored the hardware design of lattice-based FALCON signature scheme.

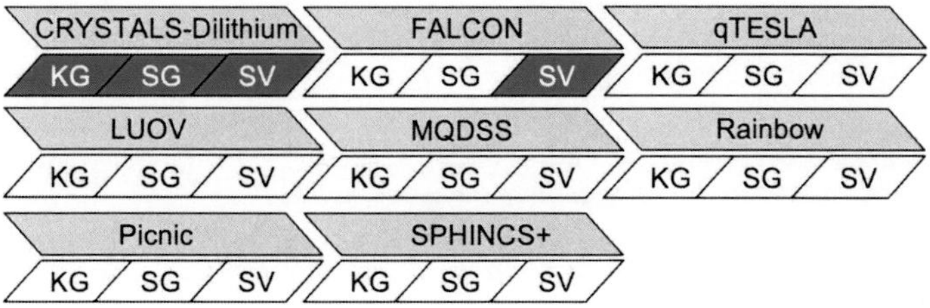

References

1. T. Prest, P.-A. Fouque, J. Hoffstein, P. Kirchner, V. Lyubashevsky, T. Pornin, T. Ricosset, G. Seiler, W. Whyte, Z. Zhang, *Falcon: Fast-Fourier Lattice-Based Compact Signatures over ntru*. Submission to the NIST Post-Quantum Cryptography Standardization Project, 2019. https://csrcnistgov/CSRC/media/Projects/Post-Quantum-Cryptography/documents/round-2/submissions/FalconRound2.zip
2. C. Gentry, C. Peikert, V. Vaikuntanathan, Trapdoors for hard lattices and new cryptographic constructions, in *Proceedings of the fortieth annual ACM symposium on Theory of computing*, pp. 197–206, 2008
3. L. Ducas, T. Prest, Fast Fourier orthogonalization, in *Proceedings of the ACM on International Symposium on Symbolic and Algebraic Computation*, pp. 191–198, 2016
4. D. Boneh, Ö. Dagdelen, M. Fischlin, A. Lehmann, C. Schaffner, M. Zhandry, Random oracles in a quantum world, in *International Conference on the Theory and Application of Cryptology and Information Security* (Springer, 2011), pp. 41–69

Chapter 4
qTESLA

4.1 Algorithm Description

qTESLA [1] is a family of post-quantum lattice-based signature schemes, based on the decision Ring Learning With Errors (R-LWE) problem. In the past decade, significant research has been conducted in the development of qTESLA. Bai and Galbraith constructed a scheme over standard lattices—the Bai-Galbraith scheme [2]. Subsequently, others proposed a lattice-based signature scheme, TESLA, which is tightly secure based on the R-LWE problem over lattices in the random-oracle model [3]. Ring-TESLA [4] and TESLA# [5] were subsequently designed as a construction over ideal lattices and a scheme with implementation improvements, respectively.

For the second round of the NIST Competition, twelve variants of qTESLA were proposed. There are two different types of qTESLA: provably secure qTESLA and heuristic qTESLA. The provably secure qTESLA anchors on security. The selected system parameters of provably secure qTESLA correspond to an instance of the R-LWE problem. The security reduction converts the R-LWE problem into the Existential Unforgeability under Chosen-Message Attack (EUF-CMA) in the Quantum Random Oracle Model (QROM) [1]. Heuristic qTESLA focuses on performance and key size, and the parameters are chosen heuristically for efficient implementation. Tables 4.1, 4.2, 4.3, and 4.4 describe these twelve variants across four security levels.

There are three components of the qTESLA digital signature scheme: key generation, signature generation, and signature verification. What follows is an explanation of each of the aforementioned components.

Key Generation The key generation algorithm produces the secret key used for signature generation and the public key used for signature verification (Algorithm 1). The first step in this algorithm is to generate a pre-seed random number using the AES algorithm. This pre-seed random number will create the seed for

D. Soni et al., *Hardware Architectures for Post-Quantum Digital Signature Schemes*, https://doi.org/10.1007/978-3-030-57682-0_4

Table 4.1 Security parameters, signature size, public key size, and secret key size of qTESLA NIST security level 1 and level 2 heuristic variants

	qTESLA-I	qTESLA-I-s	qTESLA-II	qTESLA-II-s
NIST security level	1	1	2	2
Approach	Heuristic	Heuristic	Heuristic	Heuristic
Parameter λ	95	95	128	128
Parameter κ	256	256	256	256
Parameter n	512	512	768	768
Parameter σ	22.93	22.93	9.73	9.73
Parameter k	1	1	1	1
Parameter q	$\approx 2^{22}$	$\approx 2^{22}$	$\approx 2^{23}$	$\approx 2^{23}$
Parameter h	30	30	39	39
Parameter L_E, η_E	1586, 2.306	1586, 2.306	859, 2.264	859, 2.264
Parameter L_S, η_S	1586, 2.306	1586, 2.306	859, 2.264	859, 2.264
Parameter E	1586	1586	1718	1718
Parameter S	1586	1586	1718	1718
Parameter B	$2^{20}-1$	$2^{20}-1$	$2^{21}-1$	$2^{21}-1$
Parameter d	21	21	22	22
Parameter b_{GenA}	19	19	28	28
Parameter δ_w	0.34	0.34	0.36	0.36
Parameter δ_z	0.45	0.45	0.55	0.55
Parameter δ_{sign}	0.16	0.16	0.20	0.20
Parameter δ_{keygen}	0.63	0.63	0.23	0.23
Signature size (bytes)	1376	1568	2144	2432
Public key size (bytes)	1504	480	2336	800
Private key size (bytes)	1216	2240	1600	3136

the error polynomial e, public polynomial a, secret polynomial s, and random polynomial y. Here, secret polynomial s is represented as a coefficient vector of secret polynomials. This notation is common for all the polynomials. $Seed_s$ and the nonce construct the secret polynomial with Gaussian distribution $\mathcal{D}_\sigma$. If the first h largest coefficients of secret polynomial s are less than the bound constant for secret polynomial L_S, the secret polynomial is accepted. Otherwise, a new secret polynomial is generated from a new nonce and same $seed_s$ (Lines 5–8 in Algorithm 1). Similarly, the error polynomial is constructed using a nonce and $seed_e$ with Gaussian distribution $\mathcal{D}_\sigma$. If first h largest coefficients of the error polynomial e are less than the bound constant for error polynomial L_E, the error polynomial is accepted. Otherwise, a new error polynomial is generated with a new nonce and the same $seed_e$ (Lines 9–12 in Algorithm 1). L_S and L_E reduce the signature size and key size for the algorithm. The uniform public polynomial a is generated from $seed_a$ to compute public key t; the public key comprises of both t and the $seed_a$. For the public key and secret key, the $seed_a$ is added to generate the public polynomial. Transmitting $seed_a$ instead of public polynomial a reduces the public key size and

Table 4.2 Security parameters, signature size, public key size, and secret key size of qTESLA NIST security level 3 heuristic variants

	qTESLA-III	qTESLA-III-s
NIST security level	3	3
Approach	Heuristic	Heuristic
Parameter λ	160	160
Parameter κ	256	256
Parameter n	1024	1024
Parameter σ	10.2	10.2
Parameter k	1	1
Parameter q	$\approx 2^{23}$	$\approx 2^{23}$
Parameter h	48	48
Parameter L_E, η_E	1147, 2.344	1147, 2.344
Parameter L_S, η_S	1233, 2.519	1233, 2.519
Parameter E	1147	1147
Parameter S	1233	1233
Parameter B	$2^{21}-1$	$2^{21}-1$
Parameter d	22	22
Parameter b_{GenA}	38	38
Parameter δ_w	0.43	0.43
Parameter δ_z	0.55	0.55
Parameter δ_{sign}	0.24	0.24
Parameter δ_{keygen}	0.58	0.58
Signature size (bytes)	2848	3232
Public key size (bytes)	3104	1056
Private key size (bytes)	2368	4416

Algorithm 1 qTESLA: Key generation

Input : No Input Required
Output : Secret Key $sk = (s, e, seed_a, seed_y)$ and Public Key $pk = (seed_a, t)$

procedure KEY GENERATION ▷ crypto_sign_keypair(sk, pk)
  $nonce \leftarrow 0$
  Generation of pre-seed random number using AES ▷ randombytes
  Generation of $seed_e$, $seed_a$, $seed_s$, and $seed_y$ from pre-seed. ▷ shake128/shake256
  repeat
    $nonce \leftarrow nonce + 1$
    Generation of s from $seed_s$ with Gaussian distribution. ▷ sample_gauss_poly
  until (sum(s) < secret polynomial bound constant L_S) ▷ check_ES
  repeat.
    $nonce \leftarrow nonce + 1$
    Generation of e from $seed_e$ with Gaussian distribution. ▷ sample_gauss_poly
  until (sum(e) < error polynomial bound constant L_E) ▷ check_ES
  Generation of uniform polynomial a from $seed_a$ ▷ poly_uniform
  $t_i \leftarrow a_i s + e_i$ ▷ poly_mul & poly_add_correct
  $sk \leftarrow (s, e, seed_a, seed_y)$ ▷ encode_sk
  $pk \leftarrow (t, seed_a)$ ▷ encode_sk
  return (sk, pk)
end procedure

Table 4.3 Security parameters, signature size, public key size, and secret key size of qTESLA NIST security level 5 heuristic variants

	qTESLA-V	qTESLA-V-s	qTESLA-V-size	qTESLA-V-size-s
NIST security level	5	5	5	5
Approach	Heuristic	Heuristic	Heuristic	Heuristic
Parameter λ	225	225	256	256
Parameter κ	256	256	256	256
Parameter n	2048	2048	1536	1536
Parameter σ	10.2	10.2	10.2	10.2
Parameter k	1	1	1	1
Parameter q	$\approx 2^{24}$	$\approx 2^{24}$	$\approx 2^{25}$	$\approx 2^{25}$
Parameter h	61	61	77	77
Parameter L_E, η_E	1554, 2.489	1554, 2.489	1792, 2.282	1792, 2.282
Parameter L_S, η_S	1554, 2.489	1554, 2.489	1792, 2.282	1792, 2.282
Parameter E	1554	1554	3584	3584
Parameter S	1554	1554	3584	3584
Parameter B	$2^{22} - 1$	$2^{22} - 1$	$2^{23} - 1$	$2^{23} - 1$
Parameter d	23	23	24	24
Parameter b_{GenA}	98	98	73	73
Parameter δ_w	0.31	0.31	0.33	0.33
Parameter δ_z	0.46	0.46	0.55	0.55
Parameter δ_{sign}	0.14	0.14	0.18	0.18
Parameter δ_{keygen}	0.35	0.35	0.19	0.19
Signature size (bytes)	5920	6688	4640	4640
Public key size (bytes)	6432	2336	5024	5024
Private key size (bytes)	4672	8768	3520	3520

secret key size. The secret key contains s, e, $seed_a$, and $seed_y$. $seed_y$ is a seed used to generate random polynomial y.

Signature Generation The algorithm produces a signature and its length (Algorithm 2). First, the secret polynomial s, error polynomial e, $seed_a$, and $seed_y$ are extracted. The random number r is generated using AES for polynomial y. Hence, each signature generation requires a new random number, thus making the algorithm nondeterministic. The input message (m) is hashed and $seed_a$ is used to generate the public polynomial a. Seed$_y$, random number r, and the hashed value of message (m) are then used to generate the seed for polynomial y. This seed, along with the nonce value, creates the polynomial y. The nonce value is initialized to zero and incremented each time polynomial y is generated (Lines 2, 9 in Algorithm 2). Then the value of v is computed as $(v = ay \bmod q)$. The polynomial (v) and the hashed message $H(m)$ are used as input to the hash function in order to calculate c. The variable z is computed as $z = y + sc$. The potential signature (z, c) is verified with a security test. The security test ensures z is not leaking any information about the secret s by comparing each entry of z with rejection bound $(B - S)$ (Algorithm 2:

Table 4.4 Security parameters, signature size, public key size, and secret key size of qTESLA probably secure variants

	qTESLA-p-I	qTESLA-p-III
NIST security level	1	3
Approach	Probably secure	Probably secure
Parameter λ	95	160
Parameter κ	256	256
Parameter n	1024	2048
Parameter σ	8.5	8.5
Parameter k	4	5
Parameter q	$\approx 2^{28}$	$\approx 2^{30}$
Parameter h	25	40
Parameter L_E, η_E	554, 2.61	901, 2.65
Parameter L_S, η_S	554, 2.61	901, 2.65
Parameter E	554	901
Parameter S	554	901
Parameter B	$2^{19} - 1$	$2^{21} - 1$
Parameter d	22	24
Parameter b_{GenA}	108	180
Parameter δ_w	0.37	0.33
Parameter δ_z	0.34	0.42
Parameter δ_{sign}	0.13	0.14
Parameter δ_{keygen}	0.59	0.43
Signature size (bytes)	2592	5664
Public key size (bytes)	14880	38432
Private key size (bytes)	5224	12392

Lines 14–16). If any entry of z is more than the bound (security test fails), the potential signature (z, c) is discarded and all the steps are repeated starting from Algorithm 2: Line 9. If (z) is less than or equal to the rejection bound, the algorithm goes for a subsequent correctness test, which ensures that (c) is correct and not leaking any information about the secret s. For correctness test, $v = ay - ec$ is calculated. The polynomial (v) is checked to confirm whether it is well rounded (Algorithm 2: Line 18). The polynomial v is well rounded if $|v| < (\lfloor q \rfloor /2 - E)$ and $|[v]_L| < (2^{d-1} - E)$; where q is modulus of ring R_q, d is the rounding constant in the sentence, and E is the bound constant for error polynomial e. If the correctness test fails, the potential signature (z, c) is discarded and the algorithm repeats all the steps starting from Algorithm 2: Line 9. If potential signature (z, c) passes the correctness test, the signature (z, c) is returned along with message (m) as secret message $(sm) = (z, c)$.

Signature Verification The signature verification algorithm accepts the signature if the sender is authenticated; otherwise the signature is rejected (Algorithm 3). First, it extracts (c, z) from the input signature or secret message (sm), and checks the security test of signature generation. If the test fails, the signature is rejected (Algorithm 3: Line 5, 6). The function extracts the $seed_a$ and t from the public key

Algorithm 2 qTESLA: Signature generation

Input : Message m, Message Length $mlen$, Secret Key $sk = (s, e, seed_a, seed_y)$
Output : Signature (z, c), Signature Length $smlen$

procedure SIGNATURE GENERATION ▷ crypto_sign($sm, smlen, m, mlen, sk$)
  $nonce \leftarrow 0$
  Extract $s, e, seed_a, seed_y$ from secret key ▷ decode_sk
  Generation of random number(r) using AES ▷ randombytes
  Hash the input message(m). ▷ shake128/shake256
  Generate the y-seed for polynomial y from $seed_y$, r, and $H(m)$. ▷ shake128/shake256
  Generation of uniform polynomial a from $seed_a$ ▷ poly_uniform
  while 1 **do**
    $nonce \leftarrow nonce + 1$
    Generation of polynomial y from the y-seed. ▷ sample_y
    $v \leftarrow ay \bmod q$ ▷ poly_mul
    $c \leftarrow H'(v, H(m))$ ▷ hash_H
    $z \leftarrow y + sc$ ▷ encode_c, sparse_mul16 & poly_add
    if $(z > B - S)$ **then** ▷ test_rejection
      Continue
    end if
    $v \leftarrow ay - ec$ ▷ sparse_mul & poly_sub_correct
    if (v is not well-rounded) **then** ▷ test_correctness
      Continue
    end if
    $sm \leftarrow \{(c, z), m\}$ ▷ encode_sig
    return (sm)
  end while
end procedure

Algorithm 3 qTESLA: Signature verification

Input : Signature $sm = (z, c)$, Signature length = $smlen$ Public Key $pk = (seed_a, \mathrm{t})$
Output : "Accept"/"Reject", Message m, Message Length $mlen$

procedure SIGNATURE VERIFICATION ▷ crypto_sign_open($m, mlen, sm, smlen, pk$)
  if smlen<CRYPTO_BYTES **then return** "Reject"
  end if
  $(c, z) \leftarrow sm$ ▷ decode_sig
  if abs(z) > $(B - S)$ **then return** "Reject" ▷ test_z
  end if
  $(seed_a, t) \leftarrow pk$ ▷ decode_pk
  Generation of public polynomial a using seed_a ▷ poly_uniform
  $w = az - tc$ ▷ encode_c, sparse_mul32, poly_mul & poly_sub_reduce
  Hash the input Message m ▷ shake128/shake256
  $c' \leftarrow H'(w, H(m))$ ▷ hash_H
  if $c' \neq c$ **then return** "Reject"
  end if
  $m \leftarrow sm$
  return {"Accept", m, $mlen$}
end procedure

pk. The public polynomial is generated from $seed_a$. Calculation of $w = az - tc$ is completed and the input message (m) is hashed, which is then stored in the secret message sm. Next, c', which is the hashed value of w and $H(m)$, is computed. If c' does not match with the c—generated from signature generation algorithm (Algorithm 2, it serves as an indicator that the signature is not from an authenticated user. Hence, the algorithm is terminated and a "Reject" signal is returned. If c matches with c', it reveals that a valid signature is generated by the authenticated user with the secret key, hence the signature is verified and accepted. The "Accept" output is transmitted along with message (m) and message length $mlen$.

Takeaway Only when hash message c'—the hashed value of w and $H(m)$—matches with the similar hashed output c—produced during signature generation—is the signature verification considered to be successful.

4.2 Reference C Code ⟶ HLS-Ready C Code

qTESLA is a simple and flexible signature-based algorithm in terms of C design and implementation. It requires modest efforts to transform the original C code to one that is capable of synthesis by HLS. Changes are divided into four classes as shown in Table 4.5.

- There are a few library functions in qTESLA that cannot be converted into hardware implementations using Vivado HLS. We replace library functions with C code capable of synthesis, which replicates the same functionality. For example, we replace the C `memcpy` function that copies data from one array to another, with a loop to copy the data.
- Hardware stores data in fixed-size memory, such as registers, BRAM, SRAM, and ROM. Memory size cannot be dynamically allocated or changed. Hence, we replace the dynamic memory allocation constructs with a fixed size memory. For example, while the message size varies from 33 to 3300 bytes for the Known Answer Tests (KATs), we fix it to 3300 bytes.
- HLS tools present limitations in compiling complex pointer operations. We replace the code with some functionality that does not use pointers. For example, we change the function `kmxgauss` for successful compilation.
- The qTESLA testbench uses file operations to test the algorithm using the KATs. There is no dedicated hardware module corresponding to the file operations. Hence, we change access to file operations.

Table 4.5 qTESLA: changes to the reference C code

Class	Number of changes
Replace library function	14
Remove dynamic memory allocation	7
Modify complex pointer operation	1
Change file operation	3
Total	25

Takeaway Conversion of raw C code to HLS-ready C code requires replacing pointers, certain library functions, and ensuring memory sizes are fixed.

4.3 FPGA-Specific Implementations and Optimizations

Critical functions are identified for each of qTESLA's three components (key generation, signature generation, and signature verification). These critical functions are `cSHAKE`, `sparse_mul16`, and `sparse_mul32`, respectively. We perform loop unrolling and loop pipelining optimizations on these functions to improve latency. This facilitates design space exploration for qTESLA.

```
void sparse_mul16(poly prod, const int16_t *s, const uint32_t pos_list[PARAM_H], const int16_t sign_list[PARAM_H])
{
  int i, j, pos;
  int16_t *t = (int16_t*)s;
  for (i=0; i<PARAM_N; i++)
    prod[i] = 0;
  for (i=0; i<PARAM_H; i++) {
    sparse_mul16_label2:for (j=0; j<PARAM_N; j++) {
#pragma HLS PIPELINE
#pragma HLS UNROLL factor=128
      if(j<pos_list[i])
        prod[j] = prod[j] - sign_list[i]*t[j+PARAM_N-pos_list[i]];
      else
            prod[j] = prod[j] + sign_list[i]*t[j-pos_list[i]];
    }
  }
}
```

Listing 4.1 Code: critical function of qTESLA signature generation

Listing 4.1 shows the changes in `sparse_mul16`. Vivado HLS performs better optimizations in the case when (i) there is no logic outside the inner loop, and (ii) the number of iterations for the inner and outer loops are fixed. To achieve this, the inner loop is converted into a fixed-size loop of count $PARAM\_N$, where $PARAM\_N$ is the dimension of the polynomials, i.e., each of the polynomials a, y, s, and e has $PARAM\_N$ coefficients. `qTESLA-I`, `qTESLA-II`, `qTESLA-III`, and `qTESLA-V` have a dimension of 512, 768, 1024, and 2048, respectively. Hence, the loop unrolling factors are incremented from 1 to 128 in powers of 2. Pipelining is used to parallelize the matrix computation. The `sparse_mul32` in Listing 4.2 is also modified in a similar manner.

```
void sparse_mul32(poly prod, const int32_t *pk, const uint32_t pos_list[PARAM_H], const int16_t sign_list[PARAM_H])
{
  int i, j, pos;
  for (i=0; i<PARAM_N; i++)
    prod[i] = 0;
  sparse_mul32_label0:for (i=0; i<PARAM_H; i++) {
    //pos = pos_list[i];
    sparse_mul32_label1:for (j=0; j<PARAM_N; j++) {
#pragma HLS PIPELINE
#pragma HLS UNROLL factor=128
      if(j<pos_list[i])
        prod[j] = prod[j] - sign_list[i]*pk[j+PARAM_N-pos_list[i]];
      else
            prod[j] = prod[j] + sign_list[i]*pk[j-pos_list[i]];
    }
  }
}
```

Listing 4.2 Code: critical function of qTESLA signature verification

Tables 4.6, 4.7, and 4.8 report the area overhead and latency for key generation, signature generation, and signature verification respectively. `qTESLA-II` incurs the maximum area overhead among all three components. S Signature generation and signature verification for unoptimized `qTESLA-II` do not fit into the Artix-7 FPGA. Hence, further area optimizations are performed to fit it onto this FPGA. Loop unrolling and pipelining improve the latency. Among the two, loop unrolling incurs a higher area overhead.

From Tables 4.6, 4.7, and 4.8, it is evident that security level 2 variant `qTESLA-II` takes up more area and time than the higher security level variant, `qTESLA-V`. The reason for this anomaly lies in the software implementations of these variants that we started-off with. `qTESLA-I`, `qTESLA-III`, and `qTESLA-V`, with power-of-two cyclotomic rings, have an efficient and optimal

Table 4.6 Description: Performance, area, and security (PAS) trade-off of qTESLA key generation for Artix-7 FPGA. Vivado-HLS directive loop unrolling and loop pipelining give various design points, which have different latency and area requirements, for the four qTESLA variants (qTESLA-I, qTESLA-II, qTESLA-III, and qTESLA-V). These four variants have distinct security strengths because of different algorithmic parameters

Variant	Sec. level	Optimization	FF	LUT	BRAM	DSP	Clock (ns)	Latency *Kcycles*
qTESLA-I	1	Base	22456	108764	63	69	12.65	623
		Unroll	31491	125942	64	213	12.65	604
		Pipe	22450	108880	63	69	12.65	617
qTESLA-II	2	Base	31325	127311	82	300	14.179	18760
		Unroll	32707	129840	82	270	14.179	18655
		Pipe	173157	333958	83	4482	15.1	17612
qTESLA-III	3	Base	23393	111001	78	63	12.65	3642
		Unroll	30994	124724	78	193	12.65	3608
		Pipe	23398	111122	78	63	12.65	3609
qTESLA-V	5	Base	23820	114021	117	71	12.65	32358
		Unroll	31752	128554	117	201	12.65	32058
		Pipe	23837	114152	117	71	12.65	32040

Table 4.7 Description: Performance, area, and security (PAS) trade-off of qTESLA signature generation component for Artix-7 FPGA. Vivado-HLS directive loop unrolling and loop pipelining give various designs, which have different latency and area requirements, for four qTESLA variants (qTESLA-I, qTESLA-II, qTESLA-III, and qTESLA-V) that have distinct security strength because of different algorithmic parameters

Variant	Sec. level	Optimization	FF	LUT	BRAM	DSP	Clock (ns)	Latency *Kcycles*
qTESLA-I	1	Base	23143	110287	75	73	12.667	661
		Unroll	27004	120471	77	199	12.667	419
		Pipe	23150	110370	77	73	12.667	415
qTESLA-II	2	Base	39086	137559	106	540	14.179	3696
		Unroll	38797	137363	104	540	14.179	3696
		Pipe	39086	137559	106	540	14.179	3696
qTESLA-III	3	Base	25977	125921	85	67	12.667	1030
		Unroll	27092	128486	85	97	12.667	616
		Pipe	25984	126008	85	67	12.667	587
qTESLA-V	5	Base	26324	128265	112	75	12.667	5308
		Unroll	26892	129520	112	89	12.667	3176
		Pipe	26332	128359	112	75	12.667	2870

Table 4.8 Description: Performance, area, and security (PAS) trade-off of qTESLA signature verification component for Artix-7 FPGA. Vivado-HLS directive loop unrolling and loop pipelining give various designs, which have different latency and area requirements, for four qTESLA variants (`qTESLA-I`, `qTESLA-II`, `qTESLA-III`, and `qTESLA-V`) that have distinct security strength because of different algorithmic parameters

Variant	Sec. level	Optimization	FF	LUT	BRAM	DSP	Clock (ns)	Latency *Kcycles*
`qTESLA-I`	1	Base	17683	86095	44	81	12.667	96
		Unroll	25609	101108	45	335	12.667	65
		Pipe	17690	86160	45	81	12.667	65
`qTESLA-II`	2	Base	37840	145720	73	477	14.179	1921
		Unroll	37642	145576	71	477	14.179	1921
		Pipe	37840	145720	73	477	14.179	1921
`qTESLA-III`	3	Base	17597	84765	50	72	12.667	250
		Unroll	25804	100572	50	326	12.667	153
		Pipe	17604	84834	50	72	12.667	152
`qTESLA-V`	5	Base	17885	87858	78	84	12.666	728
		Unroll	26324	120906	78	846	12.666	357
		Pipe	17967	87963	78	84	12.666	354

NTT implementation, as compared to `qTESLA-II` with non-power-of-two cyclotomic rings. qTESLA-II does not provide an optimized software implementation, and this is the root cause for the sub-optimal hardware design (Figs. 4.1, 4.2, and 4.3).

> **Takeaway** `qTESLA-II` has non-power-of-two cyclotomic rings implementation and hence, incurs more area overhead and latency.

4.4 ASIC-Specific Implementations and Optimizations

In this section, we focus on power, performance, and area (PPA) results of the ASIC design.

The power, timing, and area overhead for key pair generation, signature generation, and signature verification are reported in Tables 4.9, 4.10, and 4.11, respectively. In Table 4.9, for a given variant, the pipelined version provides better performance: $((ClockPeriod \times Latency))$, power, and area are comparable to the baseline version, except for `qTESLA-II`. For `qTESLA-II`, the pipelined version consumes maximum power due to its very high logic gate count. This is because higher logic gate area translates to higher switching activity and subsequently

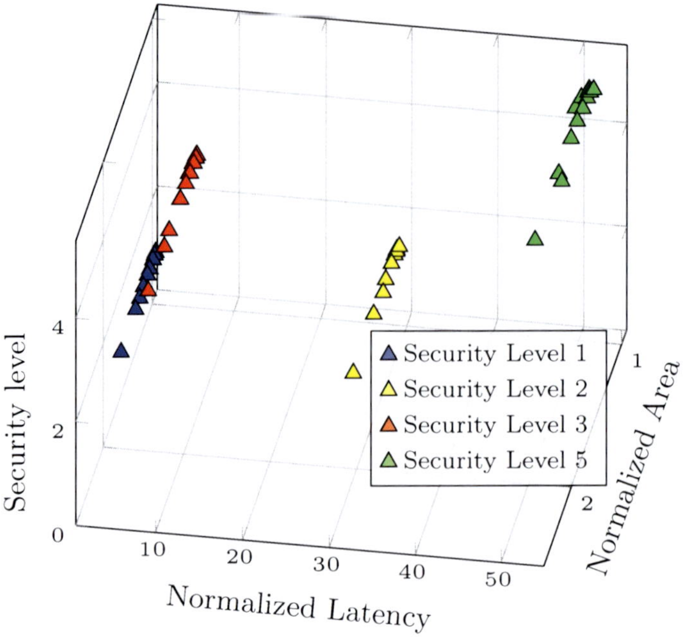

Fig. 4.1 FPGA design-space exploration of qTESLA key generation component using loop unrolling and loop pipelining optimizations. Latency and area are normalized with respect to security level 1 baseline implementation (which contains no performance optimizations)

higher power. qTESLA-II follows the same trend of more power consumption in Tables 4.10 and 4.11.

Table 4.12 shows high area utility of internal modules of qTESLA. Based on these tables, key design blocks are kmxGauss, poly_uniform, poly_mul, sparse_mul32, sample_y, hash_H, encode_c, shake256, ntt, nttinv, check_ES, KeccakF1600_StatePer, and random bytes.

kmxgauss is a Gaussian sampler function. It takes as input a nonce and a seed, and outputs a secret polynomial s or error polynomial e sampled with a Gaussian Distribution $\mathcal{D}_\sigma$. Key generation runs kmxguass multiple times until the output polynomial is bounded by L_S or L_E. Gaussian sampling (kmxgauss) is only required for qTESLA key generation operation. Key generation, signature generation, and signature verification use poly_uniform to generate a public polynomial a from input $seed_a$, which makes up part of the key pair. This approach reduces the required bandwidth as the key needs k-bits to store $seed_a$ instead of the $k * n * log_2\ q$ bits needed for full polynomial representation. poly_mul function implements polynomial multiplication over a finite field. poly_mul uses a Number Theory Transform (NTT) function such as ntt/nttinv, since qTESLA satisfies the condition $q \equiv 1(\bmod 2n)$. poly_mul, ntt, and invntt

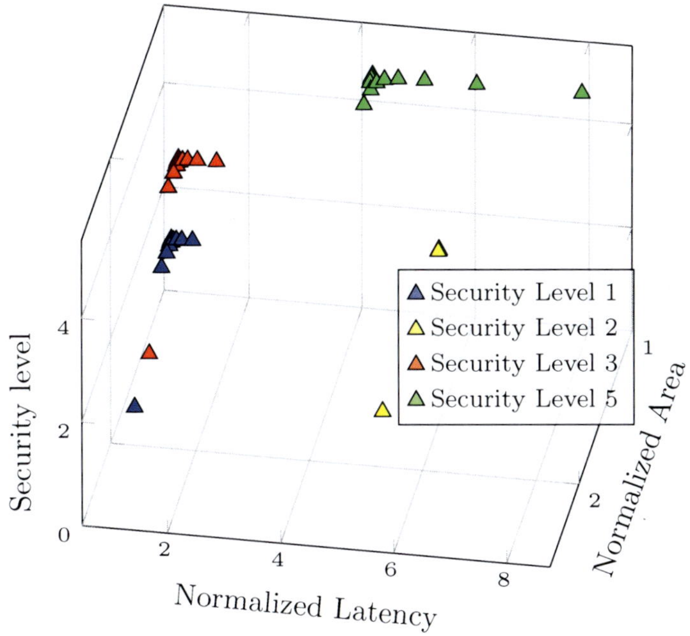

Fig. 4.2 FPGA design-space exploration of qTESLA signature generation component using loop unrolling and loop pipelining optimizations. Latency and area are normalized with respect to security level 1 baseline implementation (which contains no performance optimizations)

efficiently implement a typical polynomial multiplication. `Sample_y` uses $seed_y$ and a nonce to sample a random polynomial y, with coefficients of polynomial y in the range of $[-B, B]$. `Hash_H` is a pseudo-random function (PRF), which takes polynomial coefficients and a seed as inputs, and is used in signature generation and signature verification operations. `Encode_c` encodes the polynomial c as two arrays pos_list and $sign\_list$, where pos_list represents the position of nonzero values and $sign\_list$ embeds the sign of the nonzero value. Both output lists have h entries and provide inputs for efficient sparse multiplication. However, `sparse_mul32` implements sparse multiplication on input polynomials, which only contain h nonzero coefficients in $\{-1,1\}$; it exploits sparseness of input polynomials. `SHAKE256` function is a fundamental block of the hash function, while `keccakF1600_StatePermute` is a fundamental block for `SHAKE256` or `SHAKE128`. `check_ES` function checks upper bounds for error polynomial e and secret polynomial s. It rejects an error polynomial which has a higher bound than error polynomial bound L_E and a secret polynomial which has a higher bound than secret polynomial bound L_S. `randombytes` generates random numbers using AES.

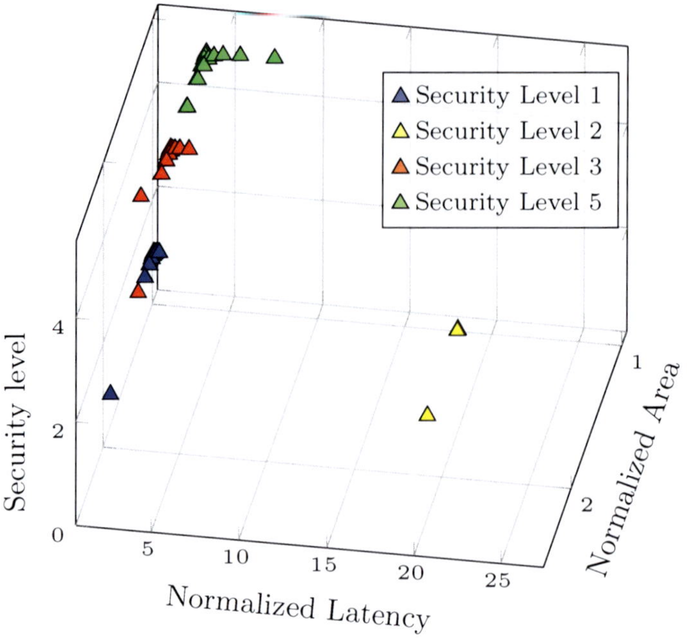

Fig. 4.3 FPGA design-space exploration of qTESLA signature verification component using loop unrolling and loop pipelining optimizations. Latency and area are normalized with respect to security level 1 baseline implementation (which contains no performance optimizations)

> **Takeaway** qTESLA-II has more logic gate area, leading to enhanced switching and hence, more power consumption.

4.5 Key Takeaways

- On the Area dimension (per security level): security level 1 < security level 3 < security level 5 < security level 2. The trend is similar for the latency dimension.
- Security level 2 exhibits sub-optimal results both in terms of area and latency. Hence, security level 2 variants should rarely be considered for any use.
- Clock period for all qTESLA schemes across all security levels is 12.65 nsec for the FPGA-based implementation (∼80 MHz), except for security level 2.
- The FPGA implementations require a larger clock period (∼12.667 nsec) as compared to ASIC-based implementations which have a clock period of ∼7 nsec. In other words, ASIC-based implementations run on a higher frequency than FPGA-based implementations of the same design.

Table 4.9 Description: ASIC-based Power, Performance, and Area (PPA) of different security levels of qTESLA key generation algorithm. Gate Count is technology independent and is derived by computing $(TotalArea - MemoryArea)$ / $(Nand2\ gate\ area)$. The Nand2 gate used to obtain the above results has an area of 2.1 μm^2 with GF (Gate Forest Technology) 65 nm Low-Power Enhanced (LPE) process technology. Given a gate count and a memory size, the ASIC area can be estimated in any technology. FF memory size in kilobytes gives the amount of memory implemented using Flip Flops instead of SRAM or SROM. Execution time can be derived by computing $(ClockPeriod \times Latency)$

Variant	Sec. level	Optimization	Power (mW)	Clock period (ns)	Area in μm^2 SRAM /ROM	Total area	Gate count $Kgate$	No of flip flops	Memory size in $Kbytes$ SRAM	SROM	FF	Latency $Kcycles$
qTESLA-I	1	Base	5.48	7.08	502349	1239040	350.81	34784	25.65	2.07	1.36	623
		Unroll	11.01	7.49	525086	1572727	498.88	43064	25.65	2.07	1.36	604
		Pipe	5.59	7.25	502349	1234242	348.52	34788	25.65	2.07	1.36	617
qTESLA-II	2	Base	12.79	7.25	701012	1930011	585.24	49417	42.66	1.75	2.07	18760
		Unroll	12.38	7.43	701012	1971268	604.88	50644	42.66	1.75	2.07	18655
		Pipe	33.31	7.71	728823	12273902	5497.66	185975	43.16	1.75	1.57	17612
qTESLA-III	3	Base	11.43	7.26	798957	1526048	346.23	35204	49.61	2.51	1.30	3642
		Unroll	12.83	7.34	798957	1814534	483.61	42205	49.61	2.51	1.30	3608
		Pipe	10.92	6.91	798957	1530645	348.42	35211	49.61	2.51	1.30	3609
qTESLA-V	5	Base	19.49	7.06	1352057	2096753	354.62	35617	95.95	5.19	1.30	32358
		Unroll	26.01	7.38	1352057	2378930	488.99	42959	95.95	5.19	1.30	32058
		Pipe	19.75	7.01	1352057	2100036	356.18	35636	95.95	5.19	1.30	32040

Table 4.10 Description: ASIC-based Power, Performance, and Area (PPA) of different security levels of qTESLA signature generation algorithm. Gate Count is technology independent and is derived by computing $(TotalArea - MemoryArea)$ / $(Nand2\ gate\ area)$. The Nand2 gate used to obtain the above results has an area of 2.1 μm^2 with GF 65 nm LPE process technology. Given a gate count and a memory size, the ASIC area can be estimated in any technology. Flip Flops memory size in kilobytes is provided in the table. Execution time can be derived by computing $(ClockPeriod \times Latency)$

Variant	Sec. level	Optimization	Power (mW)	Clock period (ns)	Area in μm^2		Gate count $Kgate$	No of flip flops	Memory size in $Kbytes$			Latency $Kcycles$
					SRAM /ROM	Total area			SRAM	SROM	FF	
qTESLA-I	1	Base	6.17	7.11	458984	1299153	400.08	43873	21.36	0.50	2.38	661
		Unroll	7.13	7.04	504459	1484319	466.60	47681	21.36	0.50	2.38	419
		Pipe	6.46	7.11	504459	1345250	400.38	43871	21.36	0.50	2.38	415
qTESLA-II	2	Base	17.62	7.45	862972	2611406	832.59	65743	55.21	0.50	3.15	3696
		Unroll	17.48	7.36	862972	2616170	834.86	65218	55.21	0.50	3.11	3696
		Pipe	17.62	7.45	862972	2611406	832.59	65743	55.21	0.50	3.15	3696
qTESLA-III	3	Base	6.21	7.18	639086	1545747	431.74	45958	41.23	0.50	2.28	1030
		Unroll	6.63	7.02	702442	1647507	450.03	46993	41.23	0.50	2.28	616
		Pipe	6.40	7.00	702442	1617091	435.55	45955	41.23	0.50	2.28	587
qTESLA-V	5	Base	13.08	7.12	1070211	1984222	435.24	46514	84.87	0.50	2.31	5308
		Unroll	13.49	7.08	1173918	2105955	443.83	47022	84.87	0.50	2.31	3176
		Pipe	13.44	7.07	1173918	2090636	436.53	46511	84.87	0.50	2.31	2870

Table 4.11 **Description:** ASIC-based Power, Performance, and Area (PPA) of different security levels of qTESLA signature verification algorithm. Gate Count is technology independent and is derived by computing $(TotalArea - MemoryArea) / (Nand2\ gate\ area)$. The Nand2 gate used to obtain the above results has an area of 2.1 μm^2 with GF 65 nm LPE process technology. Given a gate count and a memory size, the ASIC area can be estimated in any technology. Flip Flops Memory size in kilobytes is provided in the table. Execution time can be derived by computing $(ClockPeriod \times Latency)$

Variant	Sec. level	Optimization	Power (mW)	Clock period (ns)	Area in μm^2 SRAM /ROM	Area in μm^2 Total area	Gate count $Kgate$	No of flip flops	Memory size in $Kbytes$ SRAM	Memory size in $Kbytes$ SROM	Memory size in $Kbytes$ FF	Latency $Kcycles$
qTESLA-I	1	Base	4.42	7.24	282573	912573	300.00	30510	15.00	0.00	1.47	96
		Unroll	6.76	7.18	305310	1220184	435.65	38684	15.00	0.00	1.47	65
		Pipe	4.53	7.12	305310	939125	301.82	30508	15.00	0.00	1.47	65
qTESLA-II	2	Base	17.15	7.36	612253	2200392	756.26	56736	42.63	0.00	2.23	1921
		Unroll	14.95	7.37	612253	2195840	754.09	56247	42.63	0.00	2.18	1921
		Pipe	17.15	7.36	612253	2200392	756.26	56736	42.63	0.00	2.23	1921
qTESLA-III	3	Base	4.87	7.08	408624	1027650	294.77	30111	29.37	0.00	1.44	250
		Unroll	6.60	7.08	440302	1351461	433.89	38476	29.37	0.00	1.44	153
		Pipe	4.45	7.11	440302	1060638	295.40	30108	29.37	0.00	1.44	152
qTESLA-V	5	Base	11.30	6.98	771786	1402911	300.54	30677	62.01	12.50	1.46	728
		Unroll	12.68	6.87	823639	1643486	390.40	39210	62.01	12.50	1.46	357
		Pipe	11.46	6.98	823639	1453414	299.89	30675	62.01	12.50	1.46	354

Table 4.12 Description: Area consumed by the design blocks used in qTESLA

Design blocks	% of total area											
	qTESLA-I			qTESLA-II			qTESLA-III			qTESLA-V		
	Base	Unroll	Pipe	Base	Unroll	Pipe	Base	Unroll	Pipe	Base	Unroll	Pipe
Key generation												
kmxGauss	39.1	32.2	39.2	23.1	22.8	3.6	40.3	34.6	40.1	40.9	36.6	40.8
poly_uniform	25.3	19.8	25.4	17.0	16.6	2.7	22.5	19.0	22.5	22.4	19.7	22.4
ntt/nttinv	4.2	21.9	2.7	38.4	39.4	90.3	2.9	17.2	3.1	2.5	13.3	2.6
Signature generation												
poly_uniform	16.8	14.8	16.3	9.0	9.0	9.0	16.3	15.3	15.6	19.0	17.9	18.0
sample_y	14.5	12.8	14.0	7.7	7.6	7.7	18.9	17.8	18.2	16.8	15.8	16.0
hash_H	11.7	10.3	11.3	5.9	5.9	5.9	10.1	9.5	9.7	8.1	7.7	7.7
encode_c	11.1	9.8	10.7	5.7	5.7	5.7	9.7	9.1	9.3	7.9	7.4	7.5
ntt/nttinv	4.0	3.6	3.9	41.4	41.6	41.4	5.8	5.5	5.7	11.2	5.2	5.3
Signature verification												
poly_uniform	24.0	18.0	23.3	14.4	14.5	28.8	24.4	18.5	23.7	26.9	23.0	25.9
hash_H	16.6	12.5	16.2	7.0	7.1	7.0	26.4	20.1	25.6	11.5	9.9	11.1
encode_c	15.9	11.9	15.4	10.5	10.5	10.5	14.5	11.0	14.1	11.2	9.6	10.8
ntt/nttinv	5.5	27.3	5.7	54.5	54.5	54.5	4.9	25.4	4.8	6.7	17.2	6.4

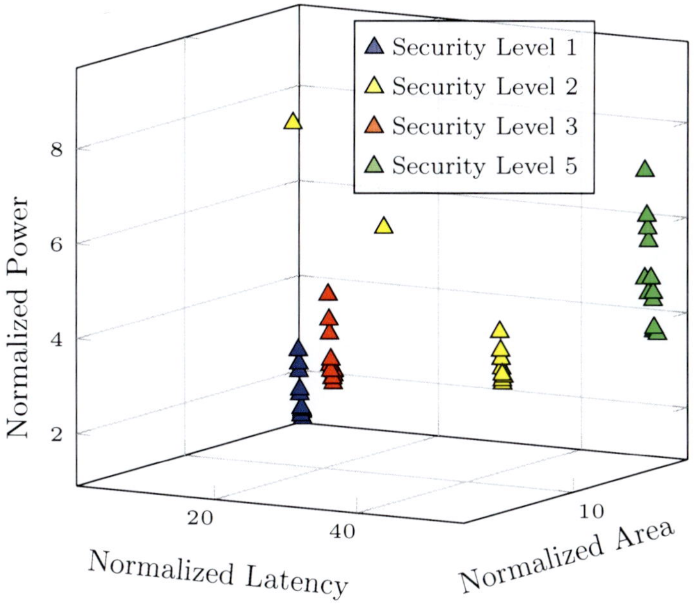

Fig. 4.4 ASIC design-space exploration of qTESLA key generation component when implemented using loop unrolling and pipelined optimizations, normalized to a baseline of security level 1 key generation

- In terms of optimization techniques, loop pipelining yields better performance. Area overhead for loop pipelining is less than that for loop unrolling and is comparable to the area overhead of the baseline. Hence, loop pipelining emerges as a better optimization technique for qTESLA.
- Power requirements differ for each security level and security level 2 consumes maximum power.
- For key generation, the latency requirement increases with security levels. Optimizations did not improve the latency even if we increase the area. Hence, baseline implementation remains the optimal option in this respect.
- For signature generation, loop unrolling and optimization improve the area and latency, thus yielding efficient design implementations.
- For signature verification, security levels 1, 3, and 5 have similar area and latency.
- Figures 4.4, 4.5, and 4.6 imply that security levels 1 and 3 have very similar PPA except for the power for key generation, where power consumption of security level 3 is on the higher side.
- For ASIC implementation, the average clock period across the variant is 7.2 nsec, which corresponds to 139 MHz. When compared to the 28 nm FPGA-based frequency numbers in Table 4.6, ASIC in 65 nm technology is 76% faster.
- The latest versions of qTESLA are `qTESLA-p-I` and `qTESLA-p-III`, which are not part of this book. We are working on hardware designs of these versions. We will share DSE of these versions on official book web page.

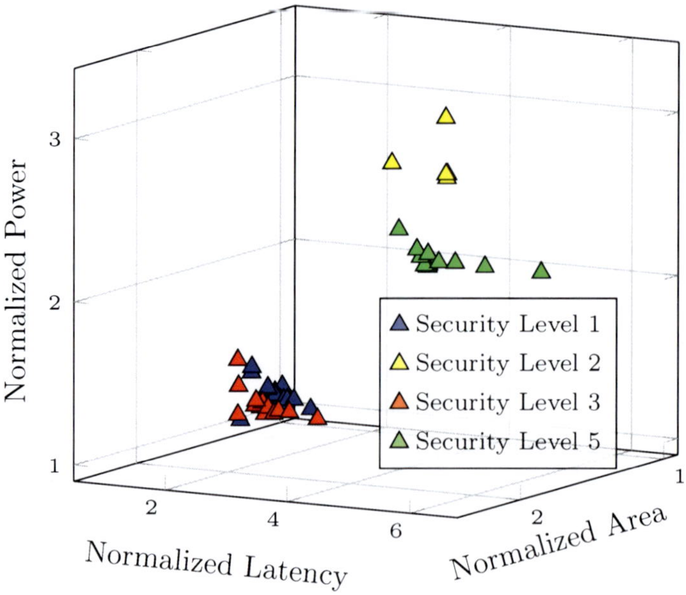

Fig. 4.5 ASIC design-space exploration of qTESLA signature generation component when implemented using loop unrolling and pipelined optimizations, normalized to a baseline of security level 1 signature generation

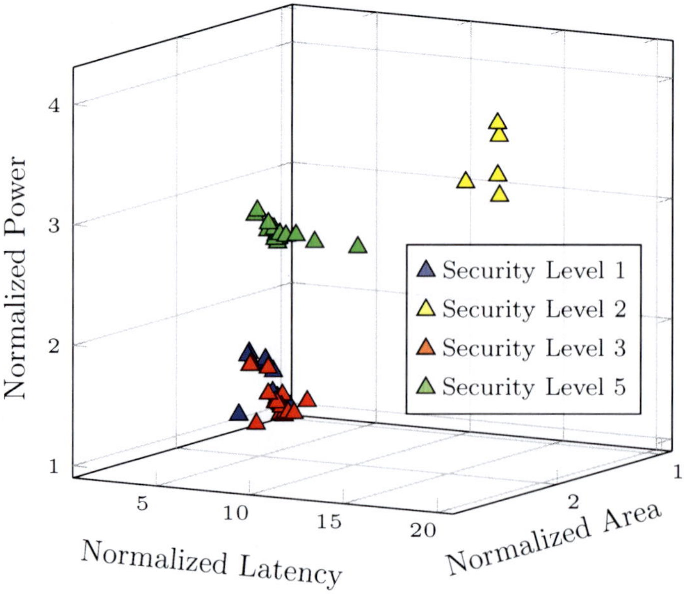

Fig. 4.6 ASIC design-space exploration of qTESLA signature verification component when implemented using loop unrolling and pipelined optimizations, normalized to a baseline of security level 1 signature verification

- With evaluation and exploration qTESLA hardware designs, we completed exploration of three lattice-based PQC signature schemes. In next chapters, we will delve into multivariate signature schemes.

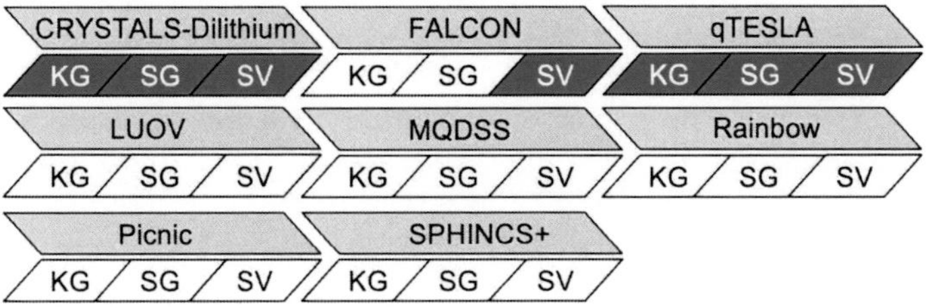

References

1. N. Bindel, S. Akleylek, E. Alkim, P.S.L.M. Barreto, J. Buchmann, E. Eaton, G. Gutoski, J. Kramer, P. Longa, H. Polat, J.E. Ricardini, G. Zanon, *Qtesla*, Submission to the NIST Post-Quantum Cryptography Standardization Project, 2019. https://csrc.nist.gov/CSRC/media/Projects/Post-Quantum-Cryptography/documents/round-2/submissions/qTESLA-Round2.zip
2. S. Bai, S.D. Galbraith, An improved compression technique for signatures based on learning with errors, in *Topics in Cryptology – CTRSA 2014*, vol. 8366, pp. 28–47, 2014.
3. E. Alkim, N. Bindel, J. Buchmann, Ö. Dagdelen, P. Schwabe, Tesla: Tightly-secure efficient signatures from standard lattices, Jul. 2015
4. S. Akleylek, N. Bindel, J. Buchmann, J. Krämer, G.A. Marson, An efficient lattice-based signature scheme with provably secure instantiation, in *Proceedings of the 8th International Conference on Progress in Cryptology AFRICACRYPT 2016*, pp. 44–60, 2016
5. P.S.L.M. Barreto, P. Longa, M. Naehrig, J.E. Ricardini, G. Zanon, *Sharper ring-lwe signatures*. Cryptology ePrint Archive, Report 2016/1026, 2016. https://eprint.iacr.org/2016/1026

Chapter 5
LUOV

5.1 Algorithm Description

The Oil and Vinegar multivariate signature scheme proposed by Patarin in 1997 [1] is one of the most analyzed multivariate schemes. With proper parameters selection, it has resisted all cryptanalytic efforts. The original scheme was broken in [2]. The Unbalanced Oil and Vinegar (UOV) cryptosystem [3] proposed some variations over the original scheme to be secure against the attack in [2]. LUOV (Lifted Unbalanced Oil and Vinegar) is a multivariate cryptography signature scheme [4, 5], designed for Existential Unforgeability under Chosen Message Attack (EUF-CMA) security, that is built upon several fundamental modifications of the UOV cryptosystems that significantly reduce the public key size. The security properties of UOV are not impacted by the adaptations made in LUOV (i.e., its security is related to the $\mathcal{MQ}$ problem). LUOV consists of six variants spanning three security levels as outlined in Tables 5.1 and 5.2.

Next, we describe the key generation, signature generation, and signature verification components of LUOV.

Key Generation The key generation component creates a secret key and a public key as shown in Algorithm 1. The first step of this component is generating the pseudo-random private seed using AES. Using this private seed, the algorithm initializes the sponge for the hash function (keccak or chacha). The public seed is generated from this hash function. The hash function also generates the matrix T ($v \times m$), where v = number of vinegar variables and m = number of polynomials in the public key = number of oil variables. T is a linear map. Q_2 is generated using the public seed and matrix T. Once the computation of both keys is complete, the public and secret keys are returned.

Signature Generation LUOV algorithm signature contains the original message. Hence, the signature generation algorithm begins by copying the input message to

D. Soni et al., *Hardware Architectures for Post-Quantum Digital Signature Schemes*, https://doi.org/10.1007/978-3-030-57682-0_5

Table 5.1 Security parameters, signature size, public key size, and secret key size of LUOV small-signature variants

	LUOV-7-57-197	LUOV-7-83-283	LUOV-7-110-374
NIST security level	1	3	5
Parameter r	7	7	7
Parameter m	57	83	110
Parameter v	197	283	374
Signature size (bytes)	239	337	440
Public key size (KB)	11.5	35.4	82
Private key size (bytes)	32	32	32

Table 5.2 Security parameters, signature size, public key size, and secret key size of LUOV signature and public key size trade-off variants

	LUOV-47-42-182	LUOV-61-60-261	LUOV-79-76-341
NIST security level	1	3	5
Parameter r	47	61	79
Parameter m	42	60	76
Parameter v	182	261	341
Signature size (bytes)	1332	2464	4134
Public key size (KB)	4.7	13.4	27.2
Private key size (bytes)	32	32	32

Algorithm 1 LUOV : key generation

Input : No Input Required
Output : Secret Key sk = (private_seed) and Public Key pk = (public_seed, Q_2)

```
1: procedure KEY GENERATION                          ▷ crypto_sign_keypair(sk, pk)
2:     Generation of private seed using AES                          ▷ randombytes
3:     Initialize the sponge.                                ▷ InitializeAndAbsorb
4:     Generate public seed using sponge.                          ▷ squeezeBytes
5:     Generate T matrix.                                 ▷ squeeze_column_array
6:     Calculate Q2.                                               ▷ calculateQ2
7:     return (sk = private_seed, pk = (public_seed, Q2))
8: end procedure
```

the signature. If the algorithm runs on partial recovery mode, the signature only stores part of the message. AES generates pseudo-random salt values to improve security against fault injection attacks and side-channel attacks. To generate the matrix T and the public seed, the algorithm follows the same flow shown in key generation, i.e., initialize the hash function and generate the public seed and matrix T from the hash function. The hash function is equipped to compute the digest of the message. This message is signed and padded with a zero byte along with a random 16-byte salt. Algorithm 2—line 8 shows that the loop runs, replacing column values and vinegar variables, trying to find a unique solution to the resulting linear system. The algorithm exits the loop when it finds a unique solution; unique solution

Algorithm 2 LUOV : signature generation

Input : Message m, Message Length $mlen$, Secret Key sk = (private_seed)
Output : Signature (m,solution,salt), Signature Length $smlen$

```
procedure SIGNATURE GENERATION                                ▷ crypto_sign(sm, smlen, m, mlen, sk)
    sig ← partial message.
    generate salt using AES and store in sig.                  ▷ randombytes
    Initialize the sponge using private seed.                  ▷ InitializeAndAbsorb
    Generate public seed using sponge.                         ▷ squeezeBytes
    Generate T matrix using sponge.                            ▷ squeeze_column_array
    Compute the target by hashing the message.                 ▷ computeTarget
    repeat
        Generate columns.                                      ▷ ColoumnGenerator_init
        Generate the vinegar variables using AES.              ▷ randombytes
        Build the augmented matrix with vinegar variables.     ▷ BuildAugmentdMatrix
        Find the unique solution for augmented matrix.         ▷ getUniqueSolution
    until (unique solution found)
    sign ← solution
    return (sign = (m, solution, salt))
end procedure
```

Algorithm 3 LUOV : signature verification

Input : Signature sm = (m,solution,salt), Signature length = $smlen$ Public Key pk = (public_seed, Q_2)
Output : "Accept"/"Reject", Message m, Message Length $mlen$

```
procedure SIGNATURE VERIFICATION                  ▷ crypto_sign_open(m, mlen, sm, smlen, pk)
    if smlen<CRYPTO_BYTES then return "Reject"
    end if
    m ← sm
    Evaluation of signature.                                   ▷ evaluatePublicMap
    Compute the target by hashing the message.                 ▷ computeTarget
    if evaluation ≠ target then return {"Reject", m, mlen}
    end if
    return {"Accept", m, mlen}
end procedure
```

discovery typically requires one iteration. The signature generation component copies the unique solution to the signature and returns this signature.

Signature verification The signature verification begins with basic signature criteria. All the signatures have a length longer than $CRYPTO\_BYTES$. $CRYPTO\_BYTES$ is the sum of the salt size and the solution size. Hence, the algorithm rejects all the signatures shorter than $CRYPTO\_BYTES$. If the signature passes the first criteria, the algorithm copies the original message in the ($message$) variable, which will be returned as the output. The algorithm evaluates the public map from the signature using the public key and stores the variable ($evaluation$). After evaluation, signature verification (similar to the signature generation) calculates the hash digest of the message that will be signed, padded with a zero byte and a random 16-byte salt. The hash digest is the *target*. If the *target* does not match the evaluation, the signature is rejected. If the target matches the evaluation, the input signature is verified.

Takeaway: In order to verify the signature, the hash digest should match with the public map from the signature using the public key.

5.2 Reference C Code ⟶ HLS-Ready C Code

Vivado HLS does not compile complex pointer operations and code structures.Therefore, we change multiple declarations, structures, and functions as shown in Table 5.3.

- LUOV uses few library functions that might be complex or might use assembly code for efficient implementation. Hence, we replace nine such functions with manual C code. We replace all the `memcpy` functions within LUOV.
- There are a few functions in LUOV which use dynamic memory allocation for the internal one-dimensional or two-dimensional arrays. We replace all memory allocations with static memory allocations.
- LUOV uses several complex pointer structures for matrix functionalities, target computation, and executing the hash function. We replace the pointer structures with corresponding code or relevant type casting.
- We change file operations in the testbench.
- LUOV signature verification does not support log and antilog functions using Look-Up Tables. We replace the corresponding functions as shown in Listings 5.1 and 5.2.
- LUOV dynamically allocates variables or structures and uses such structures within different functions. We make several modifications to the latter since the HLS tool cannot compile such structures.

Table 5.3 LUOV: changes to reference C code

Class	Number of changes
Replace a library function	9
Remove dynamic memory allocation	6
Modify complex pointer operation	5
Change file operation	3
Modify code structure	2
Total	25

Takeaway: Complex pointer structures used for matrix operations, targets computation, and executing hash functions need to be replaced with type casting.

5.3 FPGA-Specific Implementations and Optimizations

LUOV has nine variants, with security strength equivalent to security levels 2, 4 or 5. Tables 5.1 and 5.2 report the parameters of updated LUOV version 2.1. For FPGA and ASIC designs, we have implemented previous LUOV versions. Hence, the variant name and parameter values are different compared to Sect. 5.1. Table 5.1 shares the location for code of previous version. We implement three variants, one variant for each security level, and verify them using the C/RTL co-simulation.The functionality is distinct for each component of LUOV. Hence, the critical function, which occupies more latency or area, is different for each of the components. We optimize $calculateQ2$ and hash function for key generation. We optimize `Next_column` and a few "`for`" loops for signature generation. We also optimize `Next_column`, the scalar product, and `evaluatePublicMap` in signature verification. LUOV uses a hash function from the flexible XKCP family

```
#if FIELD_SIZE == 8
  prod = f8antilog(logsig[i]+logsig[j]);
#else
  prod = multiply(solution[i],solution[j]);
#endif
```

Listing 5.1 Part of original LUOV verification code

```
#if FIELD_SIZE == 8
    if(i==0 && j == 0)
      temp = f8LogTable[ONE]+f8LogTable[ONE];
  else if(i==0)
      temp = f8LogTable[ONE]+f8LogTable[sig[j-1]];
  else if(j==0)
      temp = f8LogTable[sig[i-1]]+f8LogTable[ONE];
  else
      temp = f8LogTable[sig[i-1]]+f8LogTable[sig[j-1]];
  prod = f8AntiLogTable[temp];
    //prod = f8AntiLogTable[logsig[i]+logsig[j]];
    //prod = f8antilog(logsig[i]+logsig[j]);
#else
  prod = multiply(solution[i],solution[j]);
#endif
```

Listing 5.2 Part of modified LUOV verification code

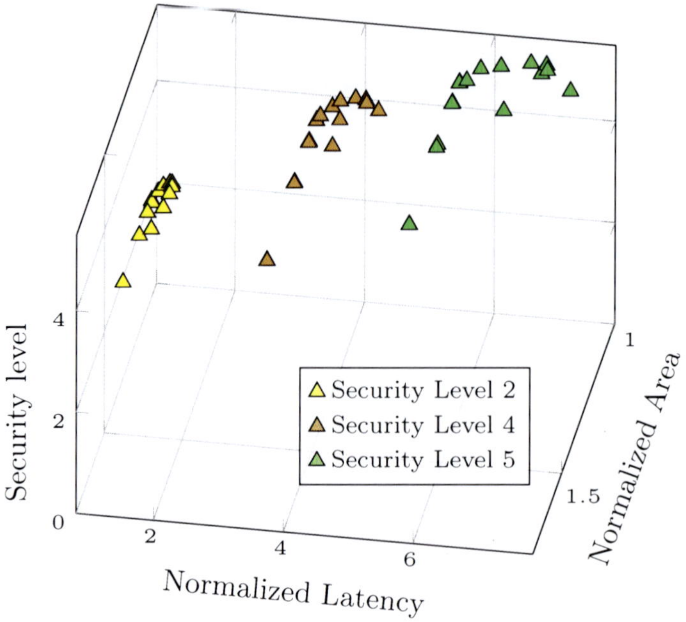

Fig. 5.1 FPGA design-space exploration of LUOV key generation component using loop unrolling and loop pipelining optimizations. Latency and area are normalized with respect to security level 1 baseline implementation (which contains no performance optimizations)

(eXtended Keccak Code Package). We also perform several optimizations on the XKCP hash functions.

Figures 5.1, 5.2, and 5.3 show that HLS optimizations improve latency for key generation and signature verification, with no significant impact on signature generation. For signature generation, the optimizations performed do not provide improved latency for functions `crypto_sign` and `Next_column`. Hence, Fig. 5.2 displays an increment in area, through parallel hardware, without a reduction in latency.

We performed different optimizations exclusively on the `calculateQ2` function to explore the design space of the key generation function. In this function, we performed loop unrolling and loop pipelining optimizations on four different loops. As `calculateQ2` is the module with the highest latency, reducing the latency of this module reduces the overall latency of this component. As a result, a fundamental area-speed curve for the key generation, presented in Fig. 5.1, is built by HLS.

Signature verification uses look-up table-based log-antilog functions for multiplication. We modified these functions, as shown in Listings 5.1 and 5.2, for creation of HLS-ready C code. We perform optimization directives on part of the code shown in Listing 5.2, which is an inner loop of `evaluatePublicMap` function. Apart from this, we also optimize loops in `Next_Column` and `_addScalarProduct`.

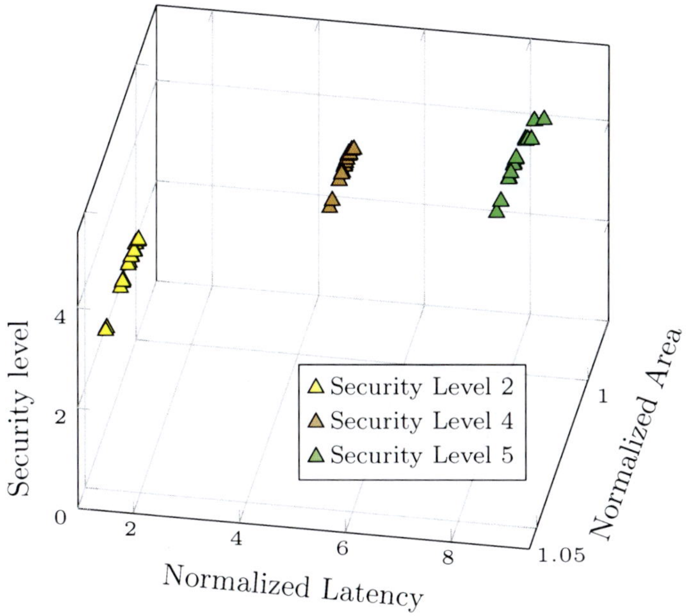

Fig. 5.2 FPGA design-space exploration of LUOV signature generation component using loop unrolling and loop pipelining optimizations. Latency and area are normalized with respect to security level 1 baseline implementation (which contains no performance optimizations)

The results of the optimizations on the signature verification function are shown in Fig. 5.3.

Based on the above optimizations, Tables 5.4, 5.5, and 5.6 report the area and performance of these designs. These designs require more LUTs and Flip-flops (FFs). Hence, they do not fit directly on the Artix-7 FPGA.

Takeaway: `calculateQ2` is the most important function for LUOV key generation, for which various optimizations are performed.

5.4 ASIC-Specific Implementations and Optimizations

In this section, we discuss the ASIC implementation results. Tables 5.7, 5.8, and 5.9 report the power, timing, and area overhead for key generation, signature generation, and signature verification, respectively.

In Table 5.7, across the security levels, the loop unrolled implementation provides a better execution time ($Latency \times ClockPeriod$), and loop unrolled

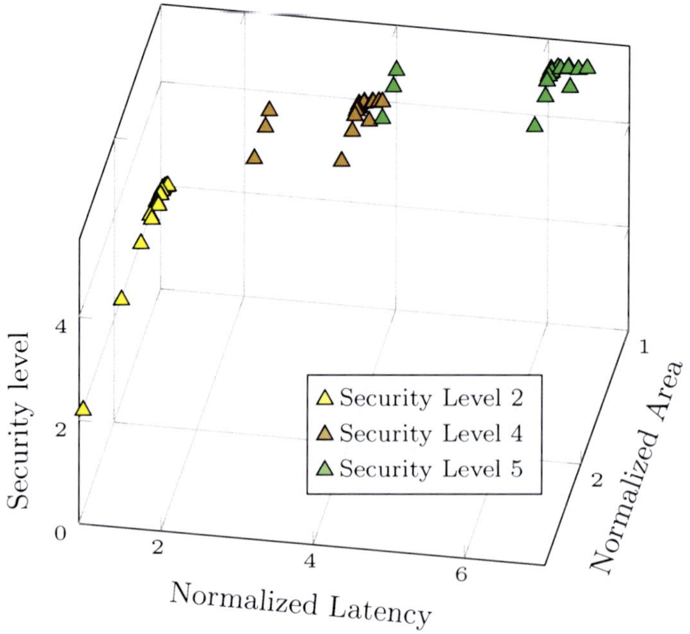

Fig. 5.3 FPGA design-space exploration of LUOV signature verification component using loop unrolling and loop pipelining optimizations. Latency and area are normalized with respect to security level 1 baseline implementation (which contains no performance optimizations)

version of security level 2 has the highest gate count and power. Though there is no significant increase in the gate count from one variant to another, the memory requirement is more when one goes from a lower security level to the higher one. Similarly, in Table 5.8, there is no significant change in the gate count across the variants and optimization, and the loop unrolled version of security level 2 has the highest gate count. Also, in Table 5.8, for security level 5, the pipelined implementation gives the best PPA and for the other two security levels, loop unrolled version offers the best execution time but at a higher power. In Table 5.9, the pipelined implementation gives a better PPA for security level 2. For security levels 4 and 5, the loop unrolled implementation gives a significant improvement in latency, hence the overall execution time. The memory requirement is much less across the variant, just a few Kbytes of SRAM and SROM are used.

Table 5.10 reports high area requirements for internal LUOV modules for key generation, signature generation, and signature verification, respectively. Based on these reports, key design blocks in LUOV are `ColumnGenerator_init`, `Next_Column`, `squeezeint64_t`, `KeccakP1600_Permute_s`, `Build AugmentedMatrix`, and `KeccakWidth1600_Spon_2`.

`ColumnGenerator_init` takes the public key and initializes a sponge object *ColumnGenerator*. `Next_Column` generates a digest from the *ColumnGenerator*. `squeezeuint64` squeezes (or outputs) a 64-bit digest from

Table 5.4 Description: Performance, area, and security (PAS) trade-off of LUOV key generation component for Artix-7 FPGA. Vivado-HLS directive loop unrolling and loop pipelining give various designs, which have different latency and area requirements, for three LUOV variants (`luov-8-58-237-keccak`, `luov-8-82-323-keccak`, and `luov-8-107-371-keccak`) that have distinct security strength because of different algorithmic parameters

Variant	Sec. level	Optimization	FF	LUT	BRAM	DSP	Clock (ns)	Latency *Kcycles*
`luov-8-58-237-keccak`	2	Base	57727	249011	70	0	8.604	5768
		Unroll	95836	293557	70	0	8.604	4742
		Pipe	57696	249145	70	0	8.604	5979
`luov-8-82-323-keccak`	4	Base	55528	249149	120	4	9.83	22220
		Unroll	60763	257856	120	13	10.204	20474
		Pipe	55901	249578	120	6	9.784	23165
`luov-8-107-371-keccak`	5	Base	55597	249359	184	4	9.83	37782
		Unroll	60801	257360	184	13	10.204	33663
		Pipe	55818	249806	184	6	9.784	39172

Table 5.5 **Description**: Performance, area, and security (PAS) trade-off of LUOV signature generation component for Artix-7 FPGA. Vivado-HLS directive loop unrolling and loop pipelining give various designs, which have different latency and area requirements, for three LUOV variants (`luov-8-58-237-keccak`, `luov-8-82-323-keccak`, and `luov-8-107-371-keccak`)

Variant	Sec. level	Optimization	FF	LUT	BRAM	DSP	Clock (ns)	Latency $K cycles$
`luov-8-58-237-keccak`	2	Base	90143	395095	116	0	8.723	9517
		Unroll	91519	400792	116	0	8.723	9199
		Pipe	90125	395275	118	0	8.723	9516
`luov-8-82-323-keccak`	4	Base	86073	393851	132	7	8.75	47522
		Unroll	86302	396901	132	7	8.75	46884
		Pipe	85948	393990	132	8	9.122	47178
`luov-8-107-371-keccak`	5	Base	86085	394302	152	5	8.723	81924
		Unroll	88796	402417	152	5	8.723	79386
		Pipe	85788	394555	152	6	9.122	80288

Table 5.6 **Description**: Performance, area, and security (PAS) trade-off of LUOV signature verification component for Artix-7 FPGA. Vivado-HLS directive loop unrolling and loop pipelining give various designs, which have different latency and area requirements, for three LUOV variants (`luov-8-58-237-keccak`, `luov-8-82-323-keccak`, and `luov-8-107-371-keccak`)

Variant	Sec. level	Optimization	FF	LUT	BRAM	DSP	Clock (ns)	Latency *Kcycles*
`luov-8-58-237-keccak`	2	Base	63678	280365	49	0	8.723	6537
		Unroll	80294	317702	63	0	8.723	6192
		Pipe	70312	298577	49	0	8.723	6158
`luov-8-82-323-keccak`	4	Base	61491	281262	49	0	8.723	24946
		Unroll	96519	353661	50	0	8.723	15923
		Pipe	61328	281282	49	0	8.723	24629
`luov-8-107-371-keccak`	5	Base	61496	281417	49	0	8.723	42559
		Unroll	97355	353834	50	0	8.723	26939
		Pipe	61284	281499	49	0	8.723	40971

Table 5.7 Description: ASIC Power, Performance, and Area (PPA) of different security levels of LUOV key generation algorithm. Gate Count is technology independent and is derived by computing $(Total\ Area\ -\ Memory\ Area)\ /\ (Nand2\ gate\ area)$. The Nand2 gate used to obtain the above results has an area of 2.1 μm^2 with GF 65 nm LPE process technology. Given a gate count and a memory size, the ASIC area can be estimated in any technology. FF Memory size in Kbytes gives the amount of memory implemented using Flip Flops instead of SRAM or SROM. Execution time can be derived by computing $(Clock Period \times Latency)$

Variant	Sec-level	Optimization	Power mW	Clock period ns	Area in μm^2 SRAM/ROM	Total area	Gate count $K gate$	No of flip flops	Memory size in Kbytes SRAM	SROM	FF	Latency $K cycles$
luov-8-82-323-keccak	2	Base	5.50	4.42	439391	2339849	905	71388	30.88	0.50	1.62	5768
		Unroll	10.35	4.39	439391	2707712	1080	98807	30.88	0.50	1.62	4742
		Pipe	5.48	4.86	439391	2339896	905	71350	30.88	0.50	1.62	5979
luov-8-107-371-keccak	4	Base	6.29	4.86	1641174	3569587	918	76431	263.27	0.50	2.51	22220
		Unroll	8.39	4.58	1641174	3640880	952	81789	263.27	0.50	2.51	20474
		Pipe	5.88	4.15	1641174	3569570	918	76715	263.27	0.50	2.51	23165
luov-8-58-237-keccak	5	Base	6.53	4.24	2379745	4337786	932	79662	392.28	0.50	2.90	37782
		Unroll	8.80	4.57	2379745	4411437	967	85018	392.28	0.50	2.90	33663
		Pipe	6.14	4.14	2379745	4336730	932	79794	392.28	0.50	2.90	39172

Table 5.8 Description: ASIC power, performance, and area (PPA) of different security levels of LUOV signature generation algorithm. Gate Count is technology independent and is derived by computing $(Total\ Area\ -\ Memory\ Area)\ /\ (Nand2\ gate\ area)$. The Nand2 gate used to obtain the above results has an area of 2.1 μm^2 with GF 65 nm LPE process technology. Given a gate count and a memory size, the ASIC area can be estimated in any technology. FF Memory size in Kbytes gives the amount of memory implemented using Flip Flops instead of SRAM or SROM. Execution time can be derived by computing $(Clock Period \times Latency)$

Variant	Sec-level	Optimization	Power mW	Clock period ns	Area in μm^2 SRAM/ROM	Total area	Gate count $Kgate$	No of flip flops	Memory size in Kbytes SRAM	SROM	FF	Latency $Kcycles$
luov-8-82-323-keccak	2	Base	6.72	4.70	620114	3540746	1391	106184	24.61	16.67	1.98	9517
		Unroll	8.14	4.56	620114	3582802	1411	107996	24.61	16.67	1.98	9199
		Pipe	6.90	4.69	650527	3570386	1390	106166	24.47	16.67	1.98	9516
luov-8-107-371-keccak	4	Base	7.28	4.55	1037648	3947526	1386	103416	52.88	16.67	2.08	47522
		Unroll	7.78	4.58	1037648	3948088	1386	103675	52.88	16.67	2.08	46884
		Pipe	7.16	4.69	1037648	3944024	1384	103289	52.88	16.67	2.08	47178
luov-8-58-237-keccak	5	Base	7.55	4.56	1394710	4312294	1389	103900	79.63	16.67	2.14	81924
		Unroll	8.27	4.69	1394710	4333231	1399	105085	79.63	16.67	2.14	79386
		Pipe	7.43	4.56	1394710	4305540	1386	103596	79.63	16.67	2.14	80288

Table 5.9 **Description**: ASIC power, performance, and area (PPA) of different security levels of LUOV signature verification algorithm. Gate Count is technology independent and is derived by computing $(Total\ Area - Memory\ Area) / (Nand2\ gate\ area)$. The Nand2 gate used to obtain the above results has an area of 2.1 μm^2 with GF 65 nm LPE process technology. Given a gate count and a memory size, the ASIC area can be estimated in any technology. FF Memory size in Kbytes gives the amount of memory implemented using Flip Flops instead of SRAM or SROM. Execution time can be derived by computing $(Clock Period \times Latency)$

Variant	Sec-level	Optimization	Power mW	Clock period ns	Area in μm^2 SRAM/ROM	Total area	Gate count $K gate$	No of flip flops	Memory size in Kbytes SRAM	SROM	FF	Latency $K cycles$
luov-8-82-323-keccak	2	Base	3.58	4.33	103098	2232471	1014	74263	1.23	5.34	1.30	6537
		Unroll	4.64	4.42	207510	2517070	1100	85442	1.23	17.81	1.30	6192
		Pipe	3.86	4.25	103098	2284307	1039	77430	1.23	5.34	1.30	6158
luov-8-107-371-keccak	4	Base	3.24	4.26	115605	2230874	1008	72729	2.14	5.06	1.41	24946
		Unroll	4.31	4.87	115605	2561941	1165	95479	2.14	5.06	1.41	15923
		Pipe	3.09	4.44	115605	2228030	1006	72543	2.14	5.06	1.41	24629
luov-8-58-237-keccak	5	Base	3.26	4.43	120771	2245444	1012	73208	2.47	5.06	1.46	42559
		Unroll	4.33	4.43	120771	2592697	1177	96686	2.47	5.06	1.46	26939
		Pipe	3.11	4.43	120771	2241740	1010	72972	2.47	5.06	1.46	40971

Table 5.10 **Description**: Area consumed by the design blocks used in LUOV

	% of total area								
	luov-8-82-323 -keccak			luov-8-107-371 -keccak			luov-8-58-237 -keccak		
Design blocks	Base	Unroll	Pipe	Base	Unroll	Pipe	Base	Unroll	Pipe
Key generation									
ColumnGenerator_init	29.9	25.8	29.9	19.1	18.8	19.1	15.7	15.5	15.6
Next_Column	15.7	13.7	15.8	10.1	10.1	10.1	8.3	8.3	8.4
squeezeuint64_t	14.9	13.1	14.9	9.9	9.9	9.9	8.1	8.1	8.1
KeccakP1600_Permute_s_1	13.8	13.2	13.8	9.1	8.9	9.1	7.5	7.3	7.4
Signature generation									
BuildAugmentedMatrix	21.2	21.0	21.0	26.5	26.5	26.5	29.6	29.5	29.6
ColumnGenerator_init	19.7	19.5	19.5	17.3	17.3	17.3	15.8	15.7	15.8
squeezeuint64_t	9.8	9.7	9.8	8.9	8.9	8.9	8.2	8.2	8.2
KeccakP1600_Permute_s_3	9.1	9.0	9.0	8.2	8.2	8.2	7.5	7.5	7.5
Signature verification									
ColumnGenerator_init	31.4	27.8	30.6	30.6	26.7	30.7	30.5	26.4	30.5
KeccakWidth1600_Spon_2	30.0	26.7	29.3	16.2	26.2	30.1	29.9	25.9	30.0
Next_Column	16.5	14.7	16.1	16.2	14.1	16.2	16.2	13.9	16.1

the sponge object. KeccakP1600_Permute_s_1 and KeccakP1600_Permute_s_3 compute the permutation part of the keccak cryptographic algorithm. BuildAugmentedMatrix forms an augmented matrix for the system. KeccakWidth1600_Spon_2 absorbs the input for the keccak cryptographic algorithm.

5.5 Key Takeaways

- The clock period for all the components is $\sim$ 8.723 ns. The frequency of LUOV is $\sim$ 115 MHz.
- Regarding the dimension of area requirements, the order is as follows: key generation < signature verification < signature generation. For execution latency, the order is: key generation < signature verification < signature generation.
- For key generation, different security levels have a comparable area overhead.
- The number of BRAMs for luov-8-107-371-keccak is $\sim$2.5$\times$ higher than the number of BRAMs for luov-8-58-237-keccak. luov-8-107-371-keccak requires $\sim$6.5$\times$ more latency compared to luov-8-58-237-keccak.
- Loop unrolling provides the best performance in terms of latency and area requirements for key generation. It improves the latency by $\sim$20% while increasing LUTs and FFs by $\sim$20% and $\sim$60%.
- Loop pipelining does not improve the latency or area for key generation.
- Figure 5.1 shows that LUOV follows the fundamental area-speed trade-off curve; as the area increases, performance improves. The parallelism of hardware or area increment is achieved by loop unrolling optimization.
- For signature generation, implementation across all security levels has a similar area overhead. luov-8-107-371-keccak requires 9$\times$ more latency compared to luov-8-58-237-keccak.
- Signature generation involves a specific algorithmic complexity. Hence, it requires around 1.5$\times$ more latency and area compared to the other two components. As reported in Table 5.5, loop unrolling marginally improves the latency while increasing the area by a limited extent. Moreover, loop pipelining yields similar latency and area overhead as the baseline implementation.
- Figure 5.2 shows an increment in the area overhead with minimum or no improvement in latency. Hence, the baseline implementation would be a better implementation for the LUOV signature generation.
- Although area overhead is similar for all security levels of LUOV signature verification, the latency for security level 4 is $\sim$7$\times$ more than the latency of security level 2.
- As shown in Table 5.6, loop unrolling optimization improves the latency by $\sim$60% with 10% increment in area overhead. However, this optimization has less improvement for the security level 2 design.

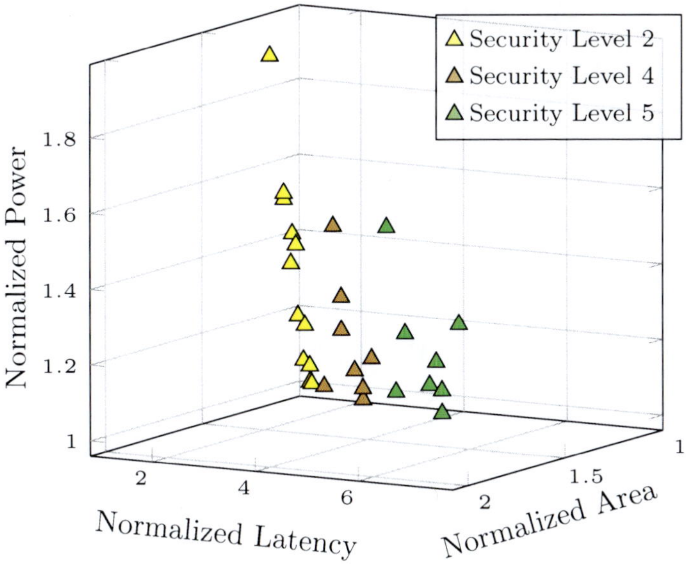

Fig. 5.4 ASIC design-space exploration of LUO key generationV component when implemented using loop unrolling and pipelined optimizations, normalized to a baseline of security level 2 key generation

- As reported in Table 5.6, loop pipelining improves latency marginally for security level 2. For other security levels, the design specifications are similar to the baseline implementations.
- Figure 5.3 reveals three optimization points for `luov-8-82-323-keccak` and `luov-8-107-371-keccak`, which have improved latency. These efficient designs directly result from adequate parallelism of public map evaluation using loop unrolling.
- As shown in Table 5.7, SRAM requirement is high for security level 4 and level 5 for the key generation algorithm compared to other variants.
- For ASIC implementation, the average clock period across the variants is 4.5 ns, which corresponds to 222 MHz. When compared to the 28 nm FPGA-based frequency numbers in Table 5.4, ASIC in 65 nm technology is around 100% faster.
- Figures 5.4, 5.5, and 5.6 show that the higher the security level, the higher the latency.
- In this chapter, we evaluated and explored the hardware design of multivariate-based LUOV signature scheme.

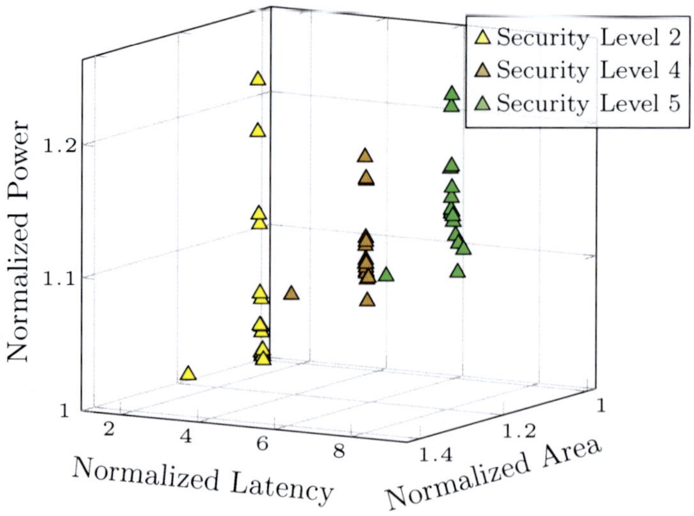

Fig. 5.5 ASIC design-space exploration of LUOV signature generation component when implemented using loop unrolling and pipelined optimizations, normalized to a baseline of security level 2 signature generation

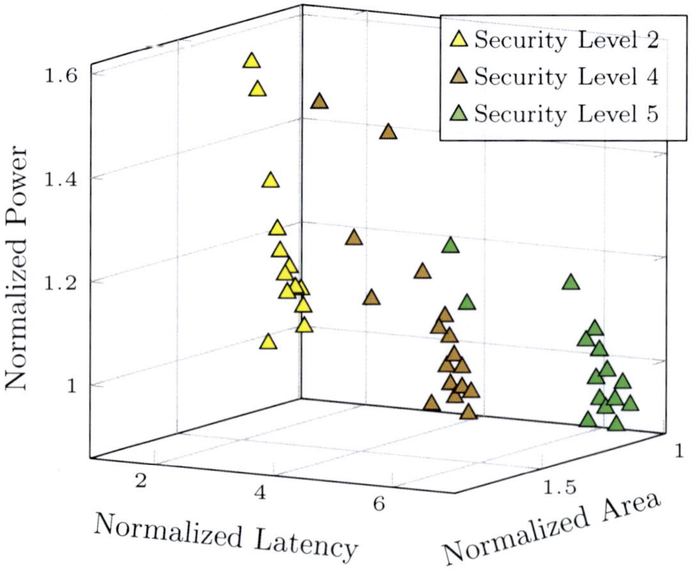

Fig. 5.6 ASIC design-space exploration of LUOV signature verification component when implemented using loop unrolling and pipelined optimizations, normalized to a baseline of security level 2 signature verification

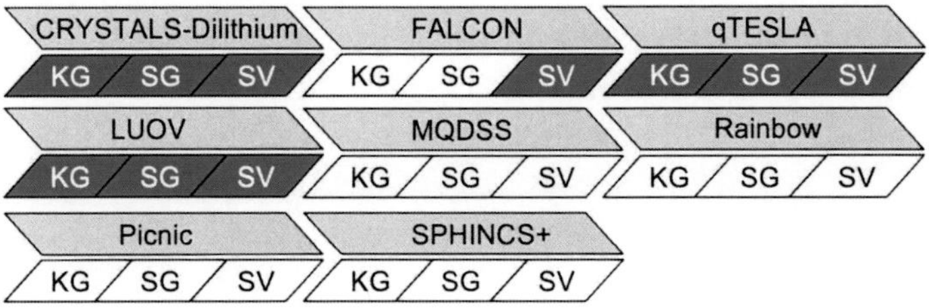

References

1. J. Patarin, The oil and vinegar signature scheme, in *Dagstuhl Workshop on Cryptography September, 1997*, 1997
2. A. Kipnis, A. Shamir, Cryptanalysis of the oil and vinegar signature scheme, in *Annual International Cryptology Conference* (Springer, New York, 1998), pp. 257–266
3. A. Kipnis, J. Patarin, L. Goubin, Unbalanced oil and vinegar signature schemes, in *International Conference on the Theory and Applications of Cryptographic Techniques* (Springer, New York, 1999), pp. 206–222
4. W. Beullens, B. Preneel, Field lifting for smaller UOV public keys, in *International Conference on Cryptology in India* (Springer, New York, 2017), pp. 227–246
5. W. Beullens, B. Preneel, A. Szepieniec, F. Vercauteren, *Luov*. Submission to the NIST Post-Quantum Cryptography Standardization Project (2019). https://csrcnistgov/CSRC/media/Projects/Post-Quantum-Cryptography/documents/round-2/submissions/LUOV-Round2.zip

Chapter 6
MQDSS

6.1 Algorithm Description

MQDSS is a digital signature scheme with a reduction from the $\mathcal{M}$ultivariate $\mathcal{Q}$uadratic ($\mathcal{MQ}$) problem [1, 2]. $\mathcal{MQ}$-systems of equations, nonlinear systems of m equations and n variables, over a finite field are hard to solve, and the decisional $\mathcal{MQ}$-problem (i.e., whether a system is solvable or not) has been proven to be $\mathcal{NP}$-complete [3]. MQDSS is secure against the Existential Unforgeability under Chosen Message Attack (EUF-CMA) and a non-tight reduction in the random oracle model [1]. MQDSS applies a Fiat-Shamir transform to the $\mathcal{MQ}$-based Sakumoto-Shirai-Hiwatari 5-pass identification scheme [4] to construct the digital signature scheme. The security of MQDSS in the Quantum Random Oracle Model (QROM) has been presented [5]. MQDSS consists of two variants spanning four security levels as outlined in Table 6.1.

The MQDSS scheme is a triplet of polynomial-time algorithms: probabilistic key generation, probabilistic signature generation, and deterministic signature verification.

Key Generation The algorithm outputs the secret key and public key pair. The algorithm flow is shown in Algorithm 1. The first step of the algorithm is to sample the secret key sk randomly using AES. Secret key sk is the input for the pseudo-random number generator which produces $skbuf$. $skbuf$ has four parts: S_F, S_s, S_ρ, and S_{rte}. Each part, in turn, acts as a seed or input for a pseudo-random function to generate different values. S_F and S_s are used to expand the system parameter F and secret input, respectively, and S_F is part of the public key's calculation. S_ρ and S_{rte} are only required for signature generation. S_F and S_s act as seeds for the hash function, which generates the inputs for the multivariate quadratic (MQ) function F and sk_{gf31} (secret input). The range of F is changed to [-15, 15] while the range of sk_{gf31} is changed to [0,31]. Using these inputs, the MQ function computes the output pk_gf31 of the nonlinear multivariate quadratic equation. The output

D. Soni et al., *Hardware Architectures for Post-Quantum Digital Signature Schemes*, https://doi.org/10.1007/978-3-030-57682-0_6

Table 6.1 Security parameters, signature size, public key size, and secret key size of MQDSS variants

	mqdss_48	mqdss_64
NIST security level	1-2	3-4
Parameter k	128	192
Parameter q	31	31
Parameter n	48	64
Parameter r	135	202
Signature size (bytes)	20854	43728
Public key size (bytes)	46	64
Private key size (bytes)	16	24

Algorithm 1 MQDSS: key generation

Input : No Input Required
Output : Secret Key $sk = (seed_{sk})$ and Public Key $pk = (S_F, pk\_gf31)$

```
procedure KEY GENERATION                                  ▷ crypto_sign_keypair(sk, pk)
    Generate sk randomly using AES.                        ▷ randombytes
    Generate expanded skbuf using sk hash.                 ▷ shake256
    Generate F and sk_gf31 using skbuf hash.               ▷ gf31_nrand_schar, gf31_nrand
    pk_gf31 ← MQ(sk_gf31, F).                              ▷ MQ
    Reduce the range of pk_gf31.                           ▷ vgf31_unique
    Concatenate pk_gf31 array.                             ▷ gf31_npack
    return (sk = (seed_sk), pk = (S_F, pk_gf31) )
end procedure
```

range of the MQ function is [0, 30]. pk_gf31 becomes part of the public key after concatenating and packing the array. Finally, the algorithm produces the secret key $sk = (seed_{sk})$ and the public key $pk = (\text{seed}_{sk}, pk\_gf31)$.

Signature Generation MQDSS signature generation signs the message m using the secret key, as presented in Algorithm 2. First, it calls MQDSS key generation using the secret key, the hash and range-limiting functions generate specific inputs—with a limited range—for the MQ. The packed output with a limited range is part of the public key. If the input has the public key, a few of the initial steps can be bypassed. The algorithm computes R, the first part of the signature, using the hash of the message and the secret key. The hash function computes D with its inputs as the public key, R—a derived message dependent random value—and the message. The inclusion of R in the signature enables signature verification to derive the same randomized digest. The signer uses a pseudo-random number generator to generate the value of ρ from S_ρ and D. The same pseudo-random number generator is also used by the signer to generate the values of r_0, t_0, and e_0 from S_{rte} and D. Next, the signer computes r_1 and gx, which are used by the commitment function com_0 and com_1 to generate the commitments c_0 and c_1. The second part of the signature σ_0 is a digest over the concatenation of all commitments. For the third part of the signature, the signer derives the challenges by applying the hash to (D, σ_0). Using the individual challenges per round, the signer computes t_1 and e_1, which provide $\sigma_2 = Packed(t_1, e_1)$. The signer computes a binary challenge b by parsing

Algorithm 2 MQDSS: signature generation

Input : Message m, Message Length $mlen$, Secret Key sk = (seed$_{sk}$)
Output : Signature $sm = (R, \sigma_0, \sigma_1, \sigma_2)$, Signature Length $smlen$

procedure SIGNATURE GENERATION ▷ crypto_sign($sm, smlen, m, mlen, sk$)
 $(S_F, S_s, S_\rho, S_{rte}) \leftarrow PRG(sk)$ ▷ `shake256`
 Generate F and sk_gf31 using S_F and S_s. ▷ $gf31\_nrand\_schar, gf31\_nrand$
 $pk\_gf31 \leftarrow MQ(sk\_gf31, F)$. ▷ `MQ`
 Reduce the range of pk_gf31. ▷ $vgf31\_unique$
 Concatenate pk_gf31 array. ▷ $gf31\_npack$
 $R \leftarrow Hash(sk, M)$. ▷ H
 $D \leftarrow Hash(pk, R, M)$. ▷ H
 $(\rho_1, \rho_2) \leftarrow (S_\rho, D)$. ▷ `shake256`
 $rnd = r_0, t_0, e_0 \leftarrow PRG(S_{rte}, D)$. ▷ $gf31\_nrand$
 for `(i=0; i < ROUNDS; i++)` **do**
  $r_1 \leftarrow (sk\_gf31, r_0)$
  $gx \leftarrow (t_0, r_1)$ ▷ G
  $gx = gx + e_0$
 end for
 for `(i=0; i < ROUNDS; i++)` **do**
  Pack r_0, t_0, e_0 ▷ $gf31\_npack$
  $c \leftarrow Hash(\rho_1, (r_0, t_0, e_0))$ ▷ com_0
  Pack and reduce the range of r_1 and gx. ▷ $vgf31\_shorten\_unique, gf31\_npack$
  $c \leftarrow Hash(\rho_2, packed(r_1, gx))$ ▷ com_1
 end for
 $\sigma_0 = Hash(c_0, c_1, ..., C_{2*ROUND-1})$. ▷ H
 $(\alpha_1, \alpha_2, ..., \alpha_3) \leftarrow hash(D, \sigma_0)$ ▷ `shake256`_absorb, `shake256`_squeezeblocks
 for `(i=0; i < ROUNDS; i++)` **do**
  $e_1 \leftarrow MQ(r_0, F)$ ▷ MQ
  $t_1 \leftarrow (\alpha, r_0, t_1)$ ▷ $vgf31\_shorten\_unique$
  $e_1 \leftarrow (\alpha, e_0, e_1)$ ▷ $vgf31\_shorten\_unique$
 end for
 $\sigma_1 \leftarrow$ Packed t_1, e_1. ▷ $gf31\_npack$
 $h1 \leftarrow Hash(D, \sigma_0, h_0, packed(t_1, e_1))$ ▷ `shake256`
 for `(i=0; i < ROUNDS; i++)` **do**
  Binary challenge(b) $\leftarrow h1$.
  $\sigma_2 \leftarrow (MUX(r_0, r_1, b), c, (\rho_1, \rho_2))$ ▷ $gf31\_npack$
 end for
 return $sm = (R, \sigma_0, \sigma_1, \sigma_2)$
end procedure

the output of hash(D, σ_0, $ch1$, σ_1). The last part of the signature consists of the concatenation of the response of the binary challenge, c, and ρ for each round. The complete signature sm = (R, Hash($c_0, c_1, ..., c_{2*ROUND-1}$), Packed($t_1$, e_1), ($resp$, c, ρ)) is the output of the signature generation algorithm.

Signature Verification As the signature verification algorithm receives a message m, a signature = $(R, \sigma_0, \sigma_1, \sigma_2)$, and a public key $pk = (S_F, pk\_gf31)$, it performs the verification flow as described in Algorithm 3. The signature consists of σ_0,

Algorithm 3 MQDSS: signature verification

Input : Signature $sm = (R, \sigma_0, \sigma_1, \sigma_2)$, Signature length = $smlen$, Public Key $pk = (S_F, pk\_gf31)$.
Output : "Accept"/"Reject", Message m, Message Length mlen

procedure SIGNATURE VERIFICATION ▷ crypto_sign_open($m, mlen, sm, smlen, pk$)
  if smlen<SIG_LEN **then return** "Reject"
  end if
  $D \leftarrow Hash(pk, R, M)$. ▷ H
  Generate F using S_F from public key. ▷ $gf31\_rand\_schar$
  $pk\_gf31 \leftarrow pk$. ▷ $gf31\_nunpack$
  $challenge_1 \leftarrow Hash(D, \sigma_0)$ ▷ `shake256_absorb`, `shake256_squeezeblocks`
  $challenge_2 \leftarrow Hash(D, \sigma_0, challenge_1, \sigma_1)$ ▷ `shake256`
  for `(i=0; i < ROUNDS; i++)` **do**
    $\alpha \leftarrow challenge_1$
    Binary challenge(b) $\leftarrow challenge_2$
    Parse σ_1 and σ_2 ▷ $gf31\_nunpack$
    if $b == 0$ **then**
      Calculate Packed r_0, t_0, e_0. ▷ MQ, $vgf31\_shorten\_unique$, $gf31\_npack$
      $c_0 \leftarrow Hash(\rho_1, (r_0, t_0, e_0))$ ▷ com_0
    else
      Calculate packed r_1, gx. ▷ MQ, G, , $vgf31\_shorten\_unique$, $gf31\_npack$
      $c_1 \leftarrow Hash(\rho_2, (r_1, gx))$. ▷ com_1
    end if
    Copy complementary commits from signature.
  end for
  $c = Hash(c_0, c_1, ..., C_{2*ROUND-1})$. ▷ H
  if c != σ_0 **then return** "Reject"
  end if
  message(m) $\leftarrow$ signature
  return {"Accept", m, $mlen$}
end procedure

which is the digest of all the commits. The signature also comprises of half of the commits in σ_2. The signature verification reconstructs the remaining commits, calculates $\sigma_0^{'}$, and verifies it with σ_0. To verify the signature, signature verification initially computes system parameter F and randomized digest message D. As the signature contains σ_0, it is simple to compute $challenge_1$ = Hash(D, σ_0) using the hash function. Similarly, the hash function computes $challenge_2$ with inputs D, parts of the signature (σ_0, σ_1), and parsed values of $challenge_1$. Binary challenge b and α are extracted from $challenge_2$ and $challenge_1$, respectively. If value b is zero, the algorithm needs to reconstruct c_0. It unpacks the values r, t, and e from the signature. Next, t_0 and e_0 are packed and computed. Using the hash function, the commitment c_0 is computed from ρ_1 (signature), r_0 (signature), packed t_0, and packed e_0. If the binary challenge b is one, commitment c_1 is calculated from inputs ρ_2 (signature), r_1, and gx. Signature verification takes the remaining required commitments from the signature. When all the rounds are completed, all the commitments are ready to compute the digest over the concatenation of all commitments. If this digest matches with σ_0, the algorithm verifies the signature. Otherwise, the algorithm rejects the signature.

Takeaway: The signature verification algorithm computes a digest over concatenation of all the commitments. Only if the digest matches σ_0, is the signature verified.

6.2 Reference C Code ⟶ HLS-Ready C Code

MQDSS uses different functions across the three components: key generation, signature generation, and signature verification. Hence, it has a very small code base with just over 1200 lines. We made several changes in the C code to make it capable of synthesis as reported in Table 6.2.

- MQDSS uses `memcpy`, `memset`, and `memmove` functions. `memset` changes the value to a constant, while `memmove` copies the data from one array to another even when there is an overlap between the arrays. We replace the library functions with functions that are capable of synthesis and have identical functionality.
- Inputs and outputs of the three components are the message, signature, public key, and secret key. We express these inputs and outputs as fixed size arrays.
- The file operations of the testbench are changed.

Takeaway: MQDSS required minimal modifications for converting the reference C code to HLS-ready C code.

6.3 FPGA-Specific Implementations and Optimizations

MQDSS algorithm has two variants: `mqdss_48` and `mqdss_64`. We implemented the hardware designs for both variants with successful verification using C/RTL co-

Table 6.2 MQDSS: changes to the reference C code to make it capable of synthesis by HLS

Class	Number of changes
Replace a library function	41
Remove dynamic memory allocation	3
Change file operation	3
Total	47

simulation. MQDSS revolves around a nonlinear multivariate quadratic equation, which requires large inputs to increase the security level of the scheme. The complex multivariate quadratic (MQ) function is critical in terms of performance and area. We applied the optimizations on both MQ and G functions.

```
void MQ(gf31 *fx, const gf31 *x, const signed char *F) {
    int i, j;
    gf31 _xij[N*(N+1) >> 1];
    int r[M] = {0};
    generate_quadratic_terms(_xij, x);
    for (i = 0; i < N; i += 2) {
        MQ_label0: for (j = 0; j < M; j++) {
            r[j] += ((int)x[i])*((int)F[i*M + 2*j]) +
                    ((int)x[i+1])*((int)F[i*M + 2*j + 1]);
    } }
    for (i = 0; i < (N*(N+1)) >> 1; i += 2) {
        MQ_label1: for (j = 0; j < M; j++) {
            r[j] += ((int)_xij[i])*((int)F[N*M + i*M + 2*j]) +
                    ((int)_xij[i+1])*((int)F[N*M + i*M + 2*j +
    1]);
    } }
    for (i = 0; i < M; i++)
        fx[i] = mod31((r[i] >> 15) + (r[i] & 0x7FFF));
}
```

Listing 6.1 Code: critical function of MQDSS

Listing 6.1 shows the MQ function which is part of the key generation, signature generation, and signature verification. This function has three loops. The first loop runs for $N \times M$ iterations; the second loop runs for $N \times (N+1)xM$ iterations, and the third loop runs for M iterations. For mqdss_48, the value of both N and M is 48. Hence, we apply the optimization on MQ_label0 of the first loop and MQ_label1 of the second loop. We use loop unrolling and pipelining for more efficient and faster hardware implementation.

Listing 6.2 shows the G function which evaluates the bilinear polar form of the MQ function, which, in essence, remains the same as the MQ function except for the missing first loop. We perform loop unrolling and loop pipelining optimization on the G_label2 loop. For key generation, we perform different optimizations on the MQ function. For signature generation and signature verification, we perform different optimizations on the MQ and G functions.

Based on the code modifications and optimization techniques, MQDSS key generation, signature generation, and signature verification hardware are generated from the C code. Figures 6.1, 6.2, and 6.3 show that these HLS optimizations improve latency for key generation, signature generation, and signature verification. The area and performance specifications are reported in Table 6.3 (for key generation),

```
void G(gf31 *fx, const gf31 *x, const gf31 *y, const signed char *F)
{
    int i, j;
    gf31 _xij[N*(N+1) >> 1];
    int r[M] = {0};
    generate_xiyj_p_xjyi_terms(_xij, x, y);
    for (i = 0; i < (N*(N+1)) >> 1; i += 2) {
        G_label2:for (j = 0; j < M; j++) {
            r[j] += ((int)_xij[i])*((int)F[N*M + i*M + 2*j]) +
                    ((int)_xij[i+1])*((int)F[N*M + i*M + 2*j + 1]);
    } }
    for (i = 0; i < M; i++)
        fx[i] = mod31((r[i] >> 15) + (r[i] & 0x7FFF));
}
```

Listing 6.2 Code: Critical function of MQDSS signature generation and verification

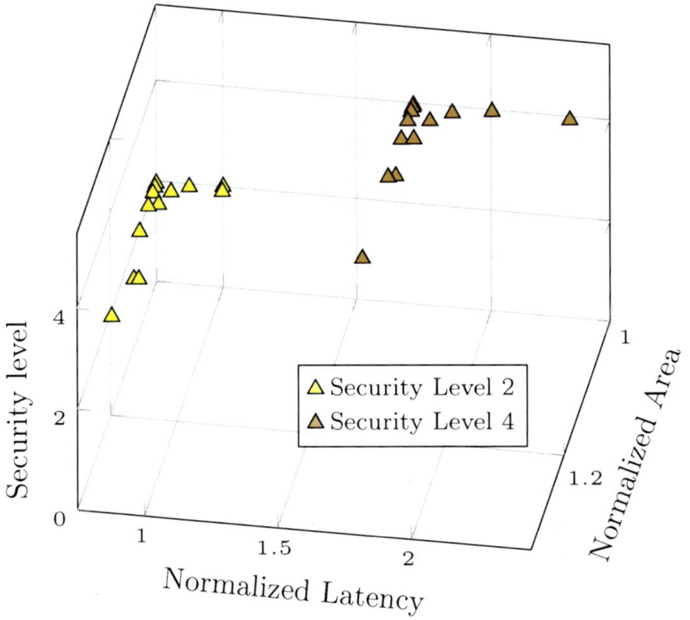

Fig. 6.1 FPGA design-space exploration of MQDSS key generation component using loop unrolling and loop pipelining optimizations. Latency and area are normalized with respect to security level 1 baseline implementation (which contains no performance optimizations)

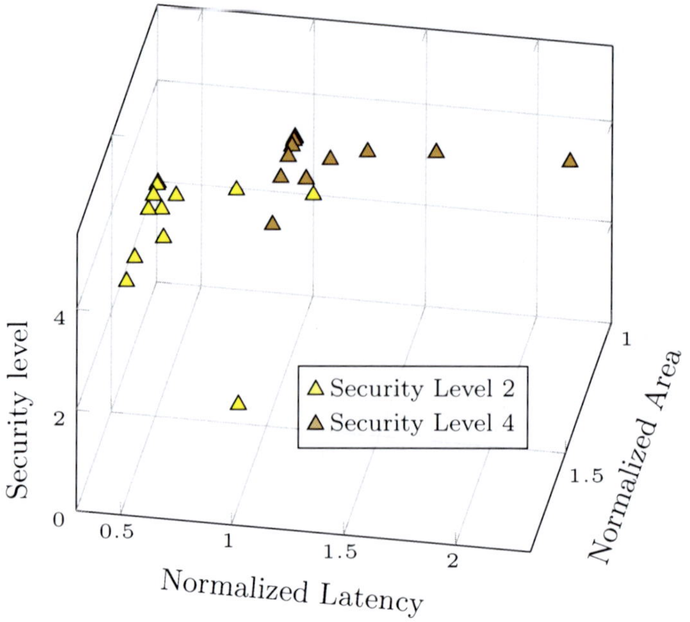

Fig. 6.2 FPGA design-space exploration of MQDSS signature generation component using loop unrolling and loop pipelining optimizations. Latency and area are normalized with respect to security level 1 baseline implementation (which contains no performance optimizations)

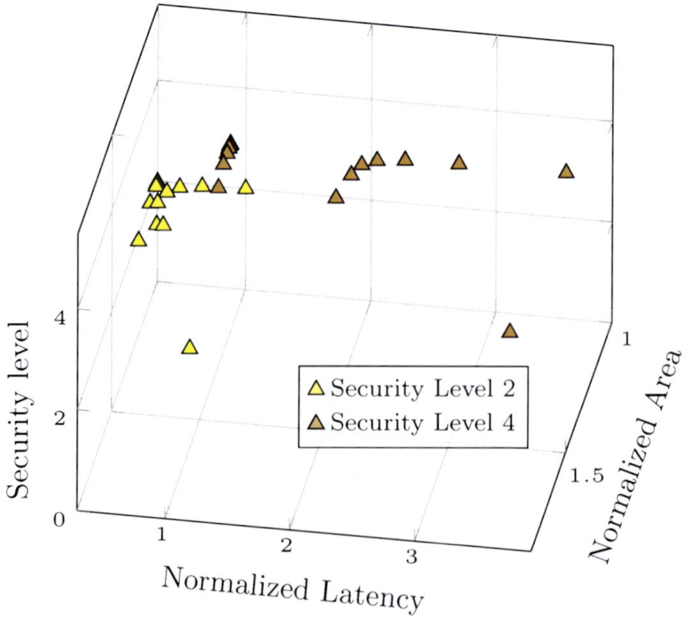

Fig. 6.3 FPGA design-space exploration of MQDSS signature verification component using loop unrolling and loop pipelining optimizations. Latency and area are normalized with respect to security level 1 baseline implementation (which contains no performance optimizations)

Table 6.3 Description: Performance, area, and security (PAS) trade-off of MQDSS key generation component for Artix-7 FPGA. Vivado-HLS directive loop unrolling and loop pipelining give various designs, which have different latency and area requirements, for two MQDSS variants (`mqdss_48` and `mqdss_64`)

Variant	Sec. level	Optimization	FF	LUT	BRAM	DSP	Clock (ns)	Latency *Kcycles*
`mqdss_48`		Base	11897	61765	58	2	7.724	364
	2	Unroll	16302	74272	59	96	8.72	276
		Pipe	12032	62001	59	2	8.295	274
`mqdss_64`		Base	11884	61703	120	2	8.334	834
	4	Unroll	17813	76082	121	128	8.72	628
		Pipe	12143	61976	121	2	8.334	625

Table 6.4 Description: Performance, area, and security (PAS) trade-off of MQDSS signature generation component for Artix-7 FPGA. Vivado-HLS directive loop unrolling and loop pipelining give various designs, which have different latency and area requirements, for two MQDSS variants (`mqdss_48` and `mqdss_64`)

Variant	Sec. level	Optimization	FF	LUT	BRAM	DSP	Clock (ns)	Latency *Kcycles*
`mqdss_48`		Base	22407	108606	170	9	8.723	34494
	2	Unroll	26707	119860	173	84	8.723	12542
		Pipe	22766	109222	173	9	8.723	10592
`mqdss_64`		Base	25274	125958	296	9	8.723	117403
	4	Unroll	27323	129940	299	44	8.723	45550
		Pipe	25931	126616	299	9	8.723	34004

Table 6.4 (for signature generation), and Table 6.5 (for signature verification). All the three components of `mqdss_48` and `mqdss_64` fit on Artix-7 FPGA. The algorithm flow for both variants are identical. `mqdss_64` has more rounds, a higher value of the system parameters (N and M), higher $HASH\_BYTES$, and higher $SEED\_BYTES$. This increases the number of computations for key generation. For the baseline implementation, the area requirement for the variants remains similar but the latency and BRAM increase significantly. Table 6.3 shows that LUT, FF, and DSP requirements are quite similar for both the variants but latency and the BRAM requirements are higher for `mqdss_64`. Moreover, the DSP requirement for loop unrolling is more than for baseline and loop pipelining.

For signature generation, there are several design points that do not fit into Artix-7; such points are not reported in Table 6.4. However, while plotting the design-space exploration graph, we consider these points for our detailed analysis. MQDSS signature generation performs key generation in the first steps. It computes the digest and all the commitments. Hence, this component has the highest latency.

Table 6.5 Description: Performance, area, and security (PAS) trade-off of MQDSS signature verification component for Artix-7 FPGA. Vivado-HLS directive loop unrolling and loop pipelining give various designs, which have different latency and area requirements, for two MQDSS variants (`mqdss_48` and `mqdss_64`)

Variant	Sec. level	Optimization	FF	LUT	BRAM	DSP	Clock (ns)	Latency *Kcycles*
`mqdss_48`		Base	17065	83847	105	6	9.945	24225
	2	Unroll	24391	103151	107	147	9.945	7416
		Pipe	17279	84191	107	6	9.945	7067
`mqdss_64`		Base	20317	101349	197	6	9.667	89948
	4	Unroll	26277	115728	198	132	9.667	48372
		Pipe	20711	101715	199	6	9.667	25010

Signature verification does not compute the public key and commitments to verify the signature. Hence, it incurs less hardware, BRAM, DSP, and latency overhead as shown in Table 6.5.

Takeaway: The MQ function is used in all three components of MQDSS. Hence, this function should be targeted for optimizations.

6.4 ASIC-Specific Implementations and Optimizations

In this section, we discuss the ASIC implementation results. Tables 6.6, 6.7, and 6.8 report the power, timing, and area overhead for key generation, signature generation, and signature verification, respectively.

In Table 6.6, for both security levels, the pipelined implementation provides a better PPA. Though the latency of the loop-unrolled and pipelined versions are comparable, the latter has a better clock period and area considerations. Also in Table 6.6, `mqdss_64` exhibits the maximum area requirements; its standard cell area (Gate Count column) is almost the same as that of `mqdss_48`, but the former has more RAM (SRAM column), hence the total area.

In Table 6.7, the pipelined version gives, overall, a better PPA compared to other optimization techniques. Although the baseline version has a slightly better area and the clock period performance, its latency is more than three times the performance of the pipelined version. Also in Table 6.7, the memory size is on the higher side. The main reason is that there are many instances of small memory that are not convertible to SRAM. Converting small memory will not give any area savings.

Table 6.6 **Description**: ASIC power, performance, and area (PPA) of different security levels of MQDSS key generation algorithm. Gate Count is technology independent and is derived by computing $(Total\ Area - Memory\ Area)\ /\ (Nand2\ gate\ area)$. The Nand2 gate used to obtain the above results has an area of 2.1 μm^2 with GF 65 nm LPE process technology. Given a gate count and a memory size, the ASIC area can be estimated in any technology. FF Memory size in kilobytes gives the amount of memory implemented using Flip Flops instead of SRAM or SROM. Execution time can be derived by computing $(Clock\ Period \times Latency)$

Variant	Sec-level	Optimization	Power mW	Clock period ns	Area in μm^2 SRAM/ROM	Total area	Gate count $K gate$	No of flip flops	Memory size in kilobytes SRAM	SROM	FF	Latency $K cycles$
mqdss_48	2	Base	2.25	3.45	666938.13	1080536.37	196.95	23573	37.78	0.00	1.31	364
		Unroll	2.48	3.57	666938.13	1203086.49	255.31	27890	37.78	0.00	1.31	276
		Pipe	2.33	3.35	666938.13	1088503.89	200.75	23610	37.78	0.00	1.31	274
mqdss_64	4	Base	2.96	3.26	1409141.94	1833983.34	202.31	24553	86.33	0.00	1.43	834
		Unroll	3.22	3.29	1409141.94	1993488.54	278.26	30359	86.33	0.00	1.43	628
		Pipe	2.93	3.35	1409141.94	1842077.22	206.16	24607	86.33	0.00	1.43	625

Table 6.7 **Description**: ASIC power, performance, and area (PPA) of different security levels of MQDSS signature generation algorithm. Gate Count is technology independent and is derived by computing $(Total\ Area - Memory\ Area)\ /\ (Nand2\ gate\ area)$. The Nand2 gate used to obtain the above results has an area of 2.1 μm^2 with GF 65 nm LPE process technology. Given a gate count and a memory size, the ASIC area can be estimated in any technology. FF Memory size in kilobytes gives the amount of memory implemented using Flip Flops instead of SRAM or SROM. Execution time can be derived by computing $(ClockPeriod \times Latency)$

Variant	Sec-level	Optimization	Power mW	Clock period ns	Area in μm^2		Gate count $Kgate$	No of flip flops	Memory size in kilobytes			Latency $Kcycles$
					SRAM/ROM	Total area			SRAM	SROM	FF	
mqdss_48	2	Base	4.26	3.79	1435255.78	2483301.46	499.07	67806	126.98	0.00	5.31	34494
		Unroll	4.84	3.83	1435255.78	2609256.82	559.05	72417	126.98	0.00	5.31	12542
		Pipe	4.30	3.81	1435255.78	2505710.74	509.74	67934	126.98	0.00	5.31	10592
mqdss_64	4	Base	5.43	4.11	2818811.40	3978544.56	552.25	74522	267.96	0.00	5.73	117403
		Unroll	5.62	4.60	2818811.40	4051994.28	587.23	76702	267.96	0.00	5.73	45550
		Pipe	5.44	3.92	2818811.40	4005412.08	565.05	74681	267.96	0.00	5.73	34004

Table 6.8 **Description**: ASIC power, performance, and area (PPA) of different security levels of MQDSS signature verification algorithm. Gate Count is technology independent and is derived by computing $(Total\ Area - Memory\ Area)\ /\ (Nand2\ gate\ area)$. The Nand2 gate used to obtain the above results has an area of 2.1 μm^2 with GF 65 nm LPE process technology. Given a gate count and a memory size, the ASIC area can be estimated in any technology. FF Memory size in kilobytes gives the amount of memory implemented using Flip Flops instead of SRAM or SROM. Execution time can be derived by computing $(ClockPeriod \times Latency)$

Variant	Sec-level	Optimization	Power mW	Clock period ns	Area in μm^2 SRAM/ROM	Total area	Gate count $Kgate$	No of flip flops	Memory size in kilobytes SRAM	SROM	FF	Latency $Kcycles$
mqdss_48	2	Base	3.15	4.38	931823.79	1680231.03	356.38	47742	75.18	0.00	3.62	24225
		Unroll	3.39	4.30	931823.79	1862076.74	442.98	54937	75.18	0.00	3.62	7416
		Pipe	3.18	4.34	931823.79	1692061.35	362.02	47824	75.18	0.00	3.62	7067
mqdss_64	4	Base	4.12	4.32	1924372.58	2785773.38	410.19	54059	165.19	0.00	3.97	89948
		Unroll	4.43	4.21	1924372.58	2944124.42	485.60	59895	165.19	0.00	3.97	48372
		Pipe	4.14	4.14	1924372.58	2802623.54	418.21	54158	165.19	0.00	3.97	25010

Table 6.9 Description: Area consumed by the design blocks used within MQDSS

Design blocks	% of total area					
	mqdss_48			mqdss_64		
	Base	Unroll	Pipe	Base	Unroll	Pipe
Key generation						
shake_256	12.1	10.9	12.0	7.2	6.6	7.1
gf31_nrand	23.8	21.4	23.6	14.0	13.0	14.0
randombytes	5.0	4.5	5.0	3.0	2.7	3.0
MQ	4.4	14.0	5.2	3.3	11.0	3.8
Signature generation						
F_ram	23.1	22.0	22.9	32.9	32.3	32.7
rnd_ram	7.9	7.5	7.9	9.4	9.2	9.3
keccak_absorb	12.2	11.6	12.1	10.0	9.8	10.0
keccak_squeezeblocks	6.6	6.2	6.6	4.1	4.0	4.0
MQ	3.9	7.6	4.6	3.0	4.4	3.6
Signature verification						
F_ram	34.1	30.8	33.9	47.0	44.4	46.7
sig_ram	8.0	7.2	7.9	9.7	9.1	9.6
keccak_absorb	18.0	16.2	17.8	14.4	13.6	14.4
keccak_squeezeblocks	4.9	4.4	4.8	2.9	2.8	2.9
MQ	2.8	9.1	3.3	2.2	7.5	2.5

In Table 6.8, the pipelined implementation gives a better PPA. The baseline version has a better area for security level 4; however, its latency is over three times that of the pipelined version, implying that the overall execution time ($LatencyXClockPeriod$) is better for the pipelined version. Additionally, the pipelined version has better power requirements for security level 4 and almost the same power for security level 2, when compared with the baseline version.

Across variants, the RAM size in mqdss_64 is more than double that of mqdss_48. There is no ROM (SROM column) used in MQDSS, whereas ROMs are mainly used in other designs, such as lattice-based algorithms, to store the twiddle factor for NTT. As expected, the loop-unrolled version of each variant has more area and hence more power.

Table 6.9 reports high area utilizing internal MQDSS modules for key generation, signature generation, and signature verification, respectively. Based on these reports, key design blocks in MQDSS are shake_256, $gf31\_nrand$, randombytes, MQ, F_ram, rnd_ram, keccak_absorb, and keccak_squeezeblocks.

Collision resistant hash function shake_256, a basic building block, is one of the most area consuming modules for all the three modules. $gf31\_nrand$ samples $gf31$ elements (which lie in the range [0,30]) and stores them in an array. It uses shake256_absorb and shake256_squeezeblocks functions. ranbombytes generates pseudo-random numbers using AES. The MQ function evaluates multi-quadratic equations for vector of N $gf31$ elements. F,

sig, and rnd array in `crypto_sign` or in `crypto_sign_open` have been implemented using registers in ASIC implementation. F and rnd arrays take a large area for signature generation. Similarly, sig and rnd arrays take up a large area for signature verification. `keccak_absorb` absorbs the data in the sponge for `shake256`. `keccak_squeezeblock` outputs the data from the sponge for `shake256`.

> **Takeaway:** Across the variants of MQDSS, RAM size of the higher security variant is more than double that of the lower security variant. Pipelined versions offer better execution time and similar power overhead compared to the baseline version.

6.5 Key Takeaways

- Minimum clock period for MQDSS components is 9.945 ns. MQDSS could run at a maximum frequency of ~100 MHz.
- Regarding the dimension of area requirements, the order is as follows: key generation < signature verification < signature generation. In regard to latency: key generation < signature verification < signature generation.
- For key generation, `mqdss_64` (security level 4) area overhead (LUTs and FFs) is similar to that of `mqdss_48` (security level 2). `mqdss_64` requires 2× more BRAMs. `mqdss_64` latency is 2.25× higher and the clock period is only 7% more than that of `mqdss_48`.
- Loop unrolling optimizations improve the latency by 25%, increase the LUTs and FFs by ~ 22%, and require 50× DSPs for parallel processing.
- Loop pipelining optimizations improve the latency by 26% without additional area overhead.
- Figure 6.1 shows several design points that increase the area without improving the latency. On such design points, we have applied loop pipelining and unrolling optimizations. However, it is observed that loop unrolling does not improve the performance.
- For signature generation, the security level 4 area overhead is ~20% more and BRAMs are ~75% more than at security level 2, while DSP and the clock period are similar for both security levels.
- Signature generation consumes more latency than the other two components. Hence, performance optimizations on this component are critical for faster devices. Latency for security level 4 is 3× more than for security level 2. Security level 2 is better suited for faster implementations.

- As shown in Table 6.4, loop unrolling optimizations improve the latency by $\sim 3\times$ and below 10% increment in LUTs and FFs compared to baseline implementation. The BRAM requirement is different at varying security levels as several design points are not considered for their larger area overhead.
- As reported in Table 6.4, loop pipelining improves the latency by $3.5\times$ without additional area overhead. As the latency numbers are quite large for signature generation, loop pipelining is the better optimization for this component.
- Figure 6.2 shows various design points with a large latency difference. Hence, the optimizations improve the latency significantly with a small increment in the area. All the loop pipelining design points have better performance.
- Figure 6.2 shows an outlier design point for security level 2. There is no area or performance optimization applied for this design. The relative position of this point—as compared to all the other design optimizations—shows the significant improvement of design through the optimizations.
- For signature verification, the security level 4 area overhead is ~25% more while the BRAM requirement is ~100% more than security level 2. The DSP and the clock period are similar for both security levels.
- For signature verification at security level 4, latency is $3.5\times$ more than security level 2. Security level 2 is better suited for faster implementations.
- As shown in Table 6.5, loop unrolling optimizations improve the latency by $\sim 3\times$ ~10–25% increment in the LUTs and FFs as compared to the baseline implementation. The DSP requirement is $\sim 24\times$ more for a highly parallel design.
- As reported in Table 6.5, loop pipelining optimizations improve the latency by $3.5\times$ without additional memory or area requirements. The loop pipelining presents a better optimization technique for signature verification.
- Figure 6.3 shows the latency difference in loop pipelining and loop unrolling optimizations for security level 4.
- The single outlier point in Fig. 6.3 for each security level is the signature verification implementation without any optimizations.
- The number of flip flops is more in ASIC than the FPGA equivalent. This is because all the memory is implemented using BRAM in FPGA. But in ASIC, the memory is converted to SRAM only if the conversion gives area gains; thus, smaller memory is not implemented as SRAM, but with flip flops.
- Figures 6.4, 6.5, and 6.6 show that security level 4 has higher PPA overhead compared to security level 2.
- In ASIC, the average clock period across the variants is around 4.45 ns, which corresponds to 225 MHz. When compared to 28 nm FPGA frequency numbers in Table 6.3, ASIC in 65 nm technology is around 87% faster for MQDSS designs.
- In this chapter, we evaluated and explored the hardware design of multivariate-based MQDSS signature scheme.

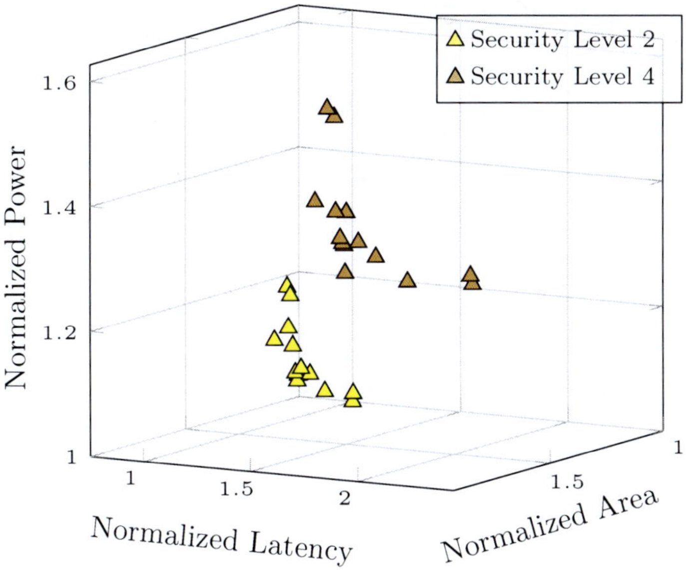

Fig. 6.4 ASIC design-space exploration of MQDSS key generation component when implemented using loop unrolling and pipelined optimizations, normalized to a baseline of security level 2 key generation

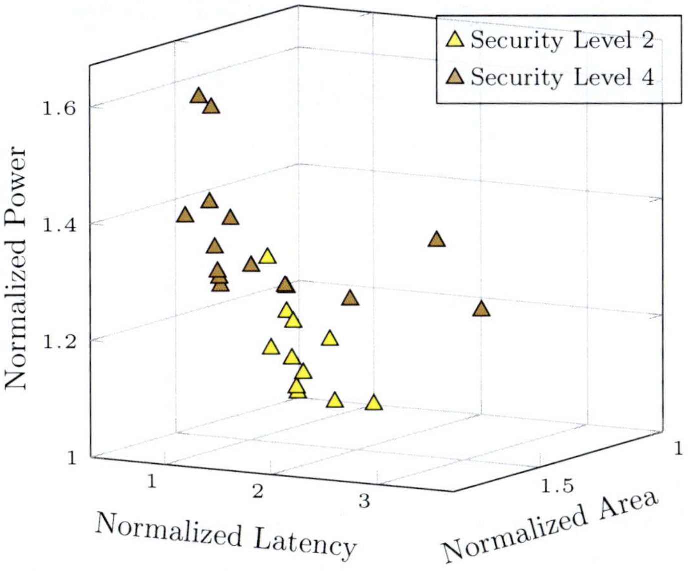

Fig. 6.5 ASIC design-space exploration of MQDSS signature generation component when implemented using loop unrolling and pipelined optimizations, normalized to a baseline of security level 2 signature generation

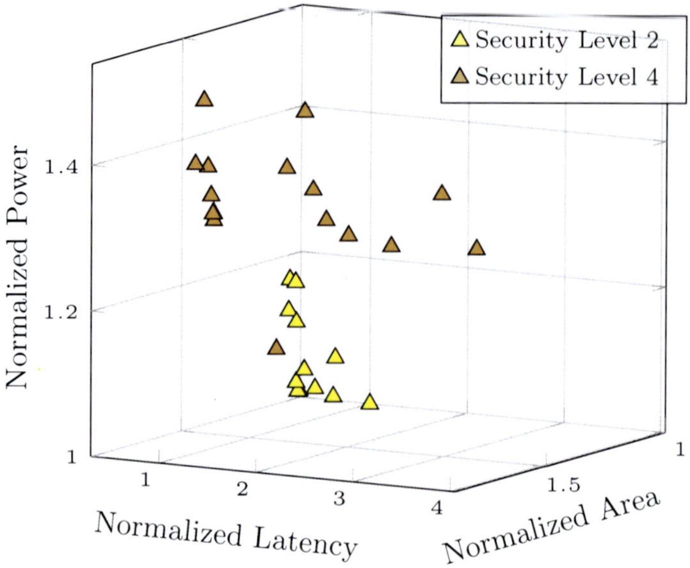

Fig. 6.6 ASIC design-space exploration of MQDSS signature verification component when implemented using loop unrolling and pipelined optimizations, normalized to a baseline of security level 2 signature verification

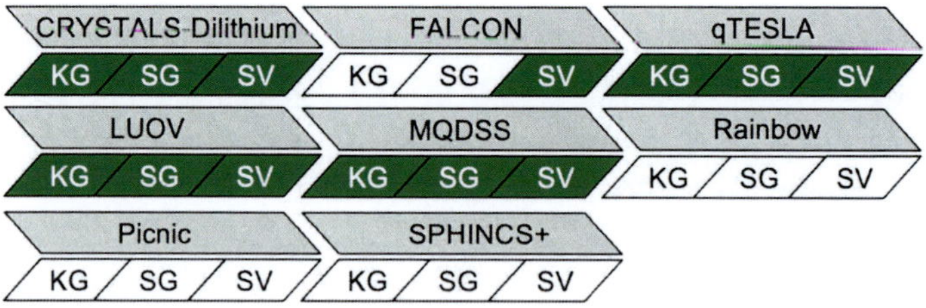

References

1. M.-S. Chen, A. Hülsing, J. Rijneveld, S. Samardjiska, P. Schwabe, From 5-pass $\mathcal{MQ}$-based identification to $\mathcal{MQ}$-based signatures, in *Advances in Cryptology – ASIACRYPT 2016*, ed. by J.H. Cheon, T. Takagi (2016), pp. 135–165
2. S. Samardjiska, M.-S. Chen, A. Hulsing, J. Rijneveld, P. Schwabe, *Mqdss*. Submission to the NIST Post-Quantum Cryptography Standardization Project (2019). https://csrcnistgov/CSRC/media/Projects/Post-Quantum-Cryptography/documents/round-2/submissions/MQDSS-Round2.zip
3. M.R. Garey, D.S. Johnson, *Computers and Intractability: A Guide to the Theory of NP-Completeness (Series of Books in the Mathematical Sciences)*, 1st edn. (W H. Freeman, New York, 1979). ISBN:0716710455

4. K. Sakumoto, T. Shirai, H. Hiwatari, Public-key identification schemes based on multivariate quadratic polynomials, in *Annual Cryptology Conference* (Springer, New York, 2011), pp. 706–723
5. J. Don, S. Fehr, C. Majenz, The measure-and-reprogram technique 2.0: multi-round fiat-shamir and more (2020). Preprint. arXiv:2003.05207

Chapter 7
Rainbow

7.1 Algorithm Description

Rainbow [1, 2] is a multivariate digital signature scheme that was proposed as an improvement to the Unbalanced Oil and Vinegar (UOV) signature scheme [3]. Such enhancement is achieved by applying the Oil and Vinegar construction multiple times in an embedded manner (i.e., the so-called layers). The security properties of UOV are not impacted by the refinements provided in Rainbow. Rainbow obtains EUF-CMA security by applying a transformation [4] and consists of three variants covering all security levels as outlined in Table 7.1. Each of these three variants has three different parameter sets: classic, cyclic [5], and compressed. In this chapter, we only consider the classic parameters.

Next, the key generation, signature generation, and signature verification algorithms of Rainbow are introduced.

Key Generation Algorithm 1 illustrates the steps for the key generation algorithm to output a public key and a secret key. The secret key includes sk_seed, S-map ($s1$), T-map ($t1, t3, t4$), and F-map ($l1\_F1, l1\_F2, l2\_F1, l2\_F2, l2\_F3, l2\_F5, l2\_F6$). The public key includes Q-map polynomials ($l1\_Q1, l1\_Q2, l1\_Q3, l1\_Q5, l1\_Q6, l1\_Q9, l2\_Q1, l2\_Q2, l2\_Q3, l2\_Q5, l2\_Q6, l2\_Q9$). The key generation process uses a PRNG instantiated by AES to generate sk_seed values, which act as input to create a secret key. A PRNG instantiated with `SHA256`generates an S-map and a T-map. This hash function also produces F-map polynomials ($layer1, layer2$). Based on this secret key, the algorithm computes an extended public key which is used for internal purposes of key generation. This extended public key has 12 parts, which are the same as the public key. `calculate_Q_from_F` computes the extended public key using the secret key. `calculate_t4` computes the $t4$ from the T-map of the secret key. Using $s1$ from the secret key, the algorithm obfuscates $layer1$ polynomials of the public key. Finally, the algorithm copies extended public key data into the public key and then returns the public key and secret key as output.

D. Soni et al., *Hardware Architectures for Post-Quantum Digital Signature Schemes*, https://doi.org/10.1007/978-3-030-57682-0_7

Table 7.1 Security parameters, signature size, public key size, and secret key size of Rainbow variants

	`Ia_Classic`	`IIIc_Classic`	`Vc_Classic`
NIST security level	1	3	5
$\mathbb{F}$	GF(16)	GF(256)	GF(256)
Parameter v_1	32	68	92
Parameter o_1	32	36	48
Parameter o_2	32	36	48
Signature size (bytes)	64	156	204
Public key size (kB)	149	710	1705.5
Private key size (kB)	93	511.4	1227.1

Algorithm 1 Rainbow: key generation

Input : No Input Required
Output : Secret Key $sk = (sk\_seed, s1, t1, t3, t4, layer1, layer2)$ and Public Key $pk = (layer1, layer2)$

procedure KEY GENERATION ▷ crypto_sign_keypair(sk, pk)
  $sk\_seed \leftarrow$ PRNG ▷ randombytes
  Generate secret key ▷ prng_set, generate_S_T, generate_B1_B2
  Calculate public key ▷ calculate_Q_from_F
  $t4 \leftarrow t1 * t3 - t4$ ▷ calculate_t4
  Obfuscation of public key ▷ obsfucate_l1_polys
  Copy the public key ▷ extcpk_to_pk
  return ($sk = (sk\_seed, s1, t3, t4, layer1, layer2)$, $pk = (layer1, layer2)$)
end procedure

Signature Generation The first step of signature generation is to generate the message digest using a hash function, as shown in Algorithm 2. Using the $digest$ and sk_seed, the algorithm generates the seed which is used to create all the pseudo-random numbers. In a loop, a pseudo-random number generator generates vinegar which is used along with $l1\_F2$ in the `gfmat_prod` function to compute layer1 matrix mat_l1. If the matrix is not solvable, the algorithm regenerates vinegar and repeats the process until it finds a solvable matrix. Once a suitable matrix is found, the algorithm proceeds to calculate variables r_l1_F1, r_l2_F1, mat_l2_F2, and mat_l2_F3 for layer2. Using these pre-calculated variables, the algorithm generates a solvable equation for layer2. In a loop, each iteration generates a new $salt$ and calculates the hash value z from $digest$ and $salt$. The hash value z and $s1$ of the secret key generate polynomial y. y, pre-computed variable r_l1_F1, and a solvable linear equation are inputs to calculate x_o1. Using r_l2_F1, mat_l2_F2, mat_l2_F3, and x_o1, the algorithm computes x_o2. If the equation is not solvable, the loop repeats the process. Eventually, the algorithm computes polynomial w and builds the signature with w and $salt$, which is the output of the signature generation process.

Algorithm 2 Rainbow: signature generation

Input : Message m, Message Length $mlen$, Secret Key $sk = (sk\_seed, s1, t1, t3, t4, layer1, layer2)$

Output : Signature $sm = (w, salt)$, Signature Length $smlen$

```
procedure SIGNATURE GENERATION                        ▷ crypto_sign(sm, smlen, m, mlen, sk)
    digest ← H(m)                                                              ▷ hash_msg
    prng_seed ← H(sk_seed, digest)                                             ▷ hash_msg
    Initialize prng                                                            ▷ prng_set
    while (!l1_succ && MAX_ATTEMPT_FRMAT > n_attempt) do
        Generating Vinegar                                                     ▷ prng_gen
        mat_l1 ← (l1_F2, vinegar)                                           ▷ gfmat_prod
        l1_succ ← linear equation solvable                                   ▷ gfmat_inv
        n_attempt ← n_attempt + 1
    end while
    Compute r_l1_F1, r_l2_F1, mat_l2_F2, mat_l2_F3               ▷ batch_quad_trimat_eval,
gfmat_prod
    while (!succ && MAX_ATTEMPT_FRMAT > n_attempt) do
        Generating salt                                                        ▷ prng_gen
        z ← H(digest, salt)                                                    ▷ hash_msg
        y ← (s1, z)                                             ▷ gfmat_prod, gf256v_add
        x_o1 ← (r_l1_F1, y)                                     ▷ gfmat_prod, gf256v_add
        x_o2 ← (r_l2_F1, r_l2_F2, mat_l2_F3, mat_l2_F2, y)                 ▷ gfmat_prod,
gf256v_add, batch_quad_trimat_eval
        Generate linear equation                                            ▷ gfmat_prod
        succ ← linear equation solvable               ▷ gfmat_prod, gf256v_add, gfmat_inv
        n_attempt ← n_attempt + 1
    end while
    w ← (y, t1, t3, t4, x_o1, x_o2)                            ▷ gfmat_prod, gf256v_add
    if MAX_ATTEMPT_FRMAT <= n_attempt then return "Reject"
    end if
    sm ← (w, salt)                                                         ▷ gf256v_add
    return sm = (w, salt)
end procedure
```

Signature Verification Algorithm 3 starts by checking the size of the signature length. If the signature length is shorter than *_SIGNATURE_BYTE*, which is the minimum length of a signature, the signature fails to verify the signer. As observed in the algorithm, the verifier computes the *digest* by hashing the message m. With the obtained *digest* and the *salt* from the signature, a hash function calculates the *correct* polynomial. The algorithm evaluates quadratic polynomials and calculates *digest_ck*. The *correct* polynomial and *digest_ck* are the same only if the signature has a correct value of the public polynomial which is based on the message and the salt. If *digest_ck* and *correct* values match, the algorithm accepts the signature and returns the message and message length. In contrast, if such values do not match, the algorithm rejects the signature.

Algorithm 3 Rainbow: signature verification

Input : Signature $sm = (w, salt)$, Signature length = $smlen$, Public Key $pk = (layer1, layer2)$.
Output : "Accept"/"Reject", Message m, Message Length $mlen$

procedure SIGNATURE VERIFICATION ▷ crypto_sign_open($m, mlen, sm, smlen, pk$)
  if smlen<_SIGNATURE_BYTE **then return** "Reject"
  end if
  $digest \leftarrow \mathrm{H}(m)$ ▷ hash_msg
  $digest\_ck \leftarrow (pk, sm)$ ▷ batch_quad_trimat_eval
  $correct \leftarrow (digest, salt)$ ▷ hash_msg
  if $digest\_ck! = correct$ **then return** "Reject"
  end if
  return {"Accept", $m, mlen$}
end procedure

7.2 Reference C Code ⟶ HLS-Ready C Code

Rainbow requires substantial changes in the C code to generate HLS-ready C code, especially for the key generation component. We divide these changes into five classes as shown in Table 7.2, which we describe here.

- Rainbow uses the OpenSSL library for hash function SHA-2. We replace `SHA256_Init`, `SHA256_Update`, and `SHA256_Final` functions with code that can be synthesized. Besides this, we also replace `memcpy`, `memset`, and `AES` library functions.
- Rainbow dynamically allocates variables and uses these variables in different functions. The HLS tool cannot compile this code structure. We make modifications to convert these dynamic memory allocations into fixed size memory allocations.
- Rainbow uses typecasting to change the datatype for the secret key, public key, and internal public key. Instead, we copy the data from one data type to another.
- We change the file operation for the testbench.
- Rainbow defines a structure for the public key and secret key. This structure has contiguous memory assigned for internal variables. To access different variables inside the structure, Rainbow assigns one pointer and adds memory location to jump to a different variable. Vivado HLS would throw an error since it cannot

Table 7.2 Rainbow: changes to reference C code

Class	Number of changes
Replacing library function	50
Removing dynamic memory allocation	10
Modifying complex pointer operation	3
Changing file operation	3
Modification of code structure	19
Total	85

access one variable with relative location to another variable of the structure. We simplify the structure to access variables.

Takeaway: It is imperative to replace the OpenSSL functions for Rainbow in order to make the code HLS-ready.

7.3 FPGA-Specific Implementations and Optimizations

Rainbow is based on nonlinear multivariate quadratic equations, and it must have large key sizes to guarantee a certain security level. Among the nine variants of Rainbow, we have implemented `Ia_Classic`, `IIIc_Classic`, and `Vc_Classic`, each of which has distinct security levels. We implemented hardware designs for these three variants with successful verification using C/RTL co-simulation.

One of the bottlenecks for HLS-ready C code conversion is the *_hash* function. As shown in Listing 7.1, the hash function uses the `SHA256` implementation from the OpenSSL library. The inputs of the hash function are *m* and *mlen*. The output is the *digest*. We replaced the three functions `SHA256_Init`, `SHA256_Update`, and `SHA256_Final` with C code that can be synthesized.

```
#include "openssl/sha.h"

static inline
int _hash(unsigned char * digest, const unsigned char * m,
    unsigned long long mlen)
{
        SHA256_CTX sha256;
        SHA256_Init(&sha256);
        SHA256_Update(&sha256, m, mlen);
        SHA256_Final(digest, &sha256);
        return 0;
}
```

Listing 7.1 Code: Hash function for Rainbow

For Rainbow variants, Tables 7.3, 7.4, and 7.5 show results of all the three components: key generation, signature generation, and signature verification, respectively. We have performed area optimizations for hash functions and performance optimizations for `rainbow_verify` and `batch_quad_trimat_eval_gf256` (or `batch_quad_trimat_eval_gf16`). Based on these optimizations, loop

Table 7.3 Description: Performance, area, and security (PAS) trade-off of Rainbow key generation for Artix-7 FPGA. Vivado-HLS loop unrolling and loop pipelining directives yield designs with different latency and area requirements, for three Rainbow variants (`Ia_Classic`, `IIIc_Classic`, and `Vc_Classic`) that have distinct security strengths because of different algorithmic parameters

Variant	Sec. level	Optimization	FF	LUT	BRAM	DSP	Clock (ns)	Latency *Kcycles*
`Ia_Classic`		Base	16237	61004	378	6	8.669	33065
	1	Unroll	17095	67743	378	6	8.669	32832
		Pipe	16337	61608	378	9	8.669	32931
`IIIc_Classic`		Base	19920	99896	1530	0	8.75	211780
	3	Unroll	28551	125847	1532	0	8.75	210519
		Pipe	20006	99989	1530	0	8.75	211018
`Vc_Classic`		Base	20720	101136	2960	0	8.75	612350
	5	Unroll	26980	127514	2960	0	8.75	608887
		Pipe	20782	101229	2960	0	8.75	610513

Table 7.4 Description: Performance, area, and security (PAS) trade-off of the Rainbow signature generation component for Artix-7 FPGA. Vivado-HLS directive loop unrolling and loop pipelining give various designs, which have different latency and area requirements, for three Rainbow variants (`Ia_Classic`, `IIIc_Classic`, and `Vc_Classic`) that have distinct security strengths because of different algorithmic parameters

Variant	Sec. level	Optimization	FF	LUT	BRAM	DSP	Clock (ns)	Latency *Kcycles*
`Ia_Classic`		Base	18725	58838	126	0	8.669	533
	1	Unroll	32091	126395	126	0	8.669	494
		Pipe	18748	59070	126	0	8.669	507
`IIIc_Classic`		Base	24304	81235	455	0	8.75	2439
	3	Unroll	35276	125185	455	0	8.75	2436
		Pipe	24355	81386	455	0	8.75	2436
`Vc_Classic`		Base	18935	79715	840	0	8.75	5723
	5	Unroll	31362	124267	840	0	8.75	5716
		Pipe	21626	88766	840	0	8.75	5337

pipelining improves the latency without significant area overhead. Tables 7.3, 7.4, and 7.5 show that the area of baseline implementation is small enough to fit into an Artix-7 FPGA. Figures 7.1, 7.2, and 7.3 display design-space exploration graphs which are normalized with baseline implementation. Baseline implementation only has area optimization related to hash functions.

Takeaway: Rainbow, which has large public key and private key, requires more latency for the key generation process compared to the signing or verification processes.

Table 7.5 Description: Performance, area, and security (PAS) trade-off of the Rainbow signature verification component for Artix-7 FPGA. Vivado-HLS directive loop unrolling and loop pipelining give various designs, which have different latency and area requirements, for three Rainbow variants (`Ia_Classic`, `IIIc_Classic`, and `Vc_Classic`) that have distinct security strength because of different algorithmic parameters

Variant	Sec. level	Optimization	FF	LUT	BRAM	DSP	Clock (ns)	Latency *Kcycles*
`Ia_Classic`		Base	5569	11223	13	0	8.669	244
	1	Unroll	14063	30974	14	0	8.674	244
		Pipe	5983	13205	13	0	8.669	156
`IIIc_Classic`		Base	9092	24676	37	0	8.75	1299
	3	Unroll	10268	27606	38	0	8.75	1298
		Pipe	9058	24743	37	0	8.75	1298
`Vc_Classic`		Base	7553	20713	29	0	8.75	3080
	5	Unroll	7839	21595	30	0	8.75	3071
		Pipe	9306	28313	29	0	8.669	1743

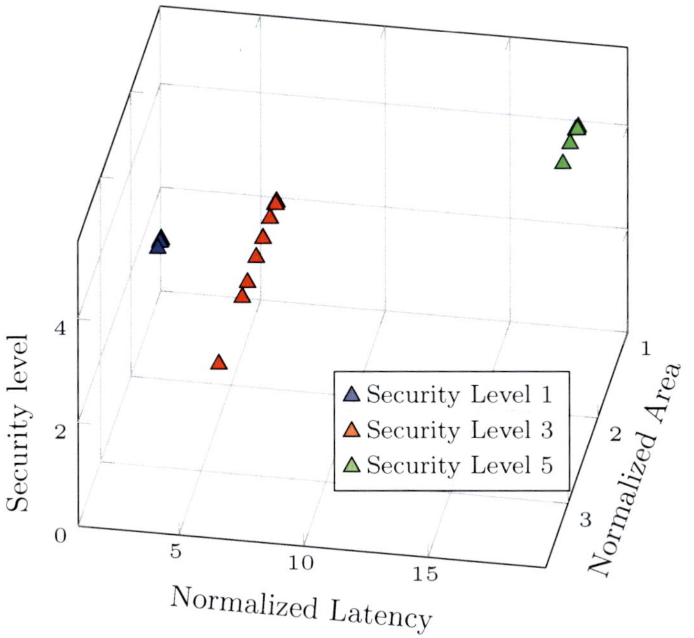

Fig. 7.1 FPGA design-space exploration of Rainbow key generation component using loop unrolling and loop pipelining optimizations. Latency and area are normalized with respect to `Ia_Classic` (i.e., security level 1) baseline implementation (which contains no performance optimizations)

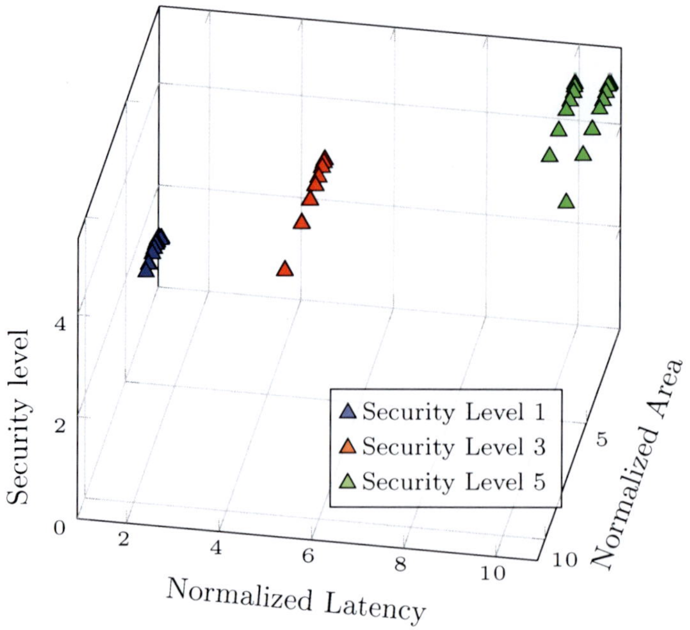

Fig. 7.2 FPGA design-space exploration of Rainbow signature generation component using loop unrolling and loop pipelining optimizations. Latency and area are normalized with respect to `Ia_Classic` (i.e., security level 1) baseline implementation (which contains no performance optimizations)

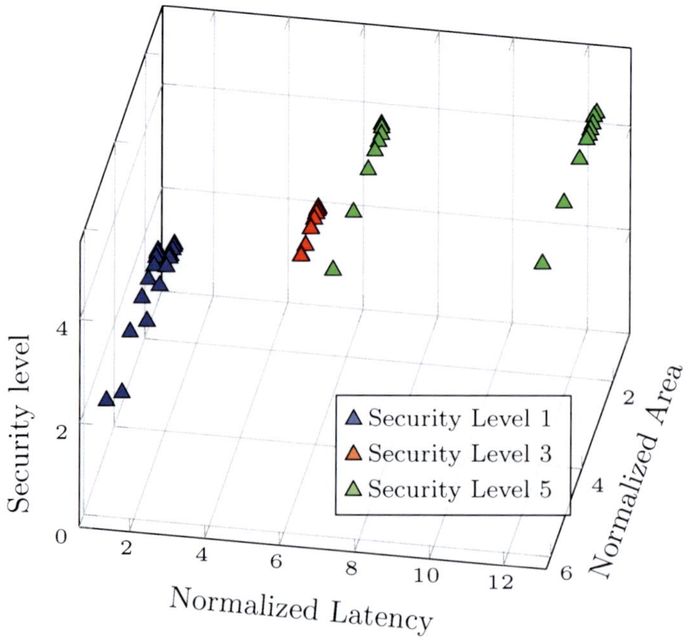

Fig. 7.3 FPGA design-space exploration of the Rainbow signature verification component using loop unrolling and loop pipelining optimizations. Latency and area are normalized with respect to `Ia_Classic` (i.e., security level 1) baseline implementation (which contains no performance optimizations)

7.4 ASIC-Specific Implementations and Optimizations

In this section, we discuss the ASIC implementation results. Tables 7.6, 7.7, and 7.8 report the power, timing, and area overhead for key generation, signature generation, and signature verification, respectively.

In Table 7.6, the loop unrolled implementation provides better throughput for security level 1, but for the other two variants, the baseline version shows a better execution time. Power and area are higher for the loop-unrolled implementation. It can be noted that although the memory size in kilobytes is the same except for in the loop unrolled version since they are implemented with multiple individual SRAM blocks, they have a higher SRAM/SROM area. In Table 7.7, memory requirement increases for higher security levels. SRAM size is around 1 MB for security level 5 and around 0.5 MB for security level 3, whereas for security level 1, it is less than 100 kilobytes. In Table 7.8, the pipelined version has better execution time across the three security levels. And the SRAM requirement is around half kilobytes, which is very less compared to key generation and signature generation algorithms.

Table 7.9 reports internal Rainbow modules with high area for key generation, signature generation, and signature verification, respectively. Based on these reports, key design blocks in Rainbow are `prng_set`, `randombytes_with_sta`, `p_hash`, `expand_hash`, and `batch_quad_trimat_ev`.

`prng_set` initializes the pseudo-random generator that uses AES. `random bytes_with_sta` takes the AES key as its input and generates a pseudo-random number. `p_hash` generates the digest for the input array. `expand_hash` runs the `p_hash` internally to calculate the hash digest for a larger array. `batch_quad_trimate_ev` evaluates the quadratic matrix.

> **Takeaway:** Across the variants of Rainbow, for key generation and signature generation, RAM size of the higher security variant is more than ten times that of the lower security variant.

7.5 Key Takeaways

- Rainbow FPGA components have a minimum clock period of 8.669 ns yielding a maximum frequency of ~115 MHz.
- In terms of area overhead and execution latency, signature verification < signature generation < key generation. Key generation requires the highest area and latency.
- In terms of latency, `Ia_Classic < IIIc_Classic < Vc_Classic`.

Table 7.6 Description: ASIC Power, Performance, and Area (PPA) of different security levels of the Rainbow key generation algorithm. Gate Count is technology independent and is derived by computing $(Total\ Area - Memory\ Area) / (Nand2\ gate\ area)$. The Nand2 gate used to obtain the above results has an area of 2.1 μm^2 with GF (Gate Forest Technology) 65 nm Low-Power Enhanced (LPE) process technology. Given a gate count and a memory size, the ASIC area can be estimated in any technology. FF Memory size in kilobytes gives the amount of memory implemented using Flip Flops instead of SRAM or SROM. Execution time can be derived by computing $(Clock Period \times Latency)$

Variant	Sec-level	Optimization	Power mW	Clock period ns	Area in μm^2 SRAM/ROM	Area in μm^2 Total area	Gate count $K gate$	No of flip flops	Memory size in Kbytes SRAM	Memory size in Kbytes SROM	Memory size in Kbytes FF	Latency $K cycles$
Ia_Classic	1	Base	9.68	4.39	4915303	5302684	184	27243	418.05	27.25	1.31	33065
		Unroll	11.13	4.00	6084747	6477847	187	28009	418.05	27.25	1.31	32832
		Pipe	9.74	4.32	4915303	5305480	186	27336	418.05	27.25	1.31	32931
IIIc_Classic	3	Base	22.25	4.35	20467996	20980400	244	34052	1936.67	64.34	1.63	211780
		Unroll	29.84	4.46	26679831	27317743	304	42202	1936.67	64.34	1.76	210519
		Pipe	22.32	4.43	20467996	20981150	244	34134	1936.67	64.34	1.63	211018
Vc_Classic	5	Base	40.67	3.95	47897328	48437900	257	35357	4636.98	122.75	1.70	612350
		Unroll	57.12	4.45	62792706	63436284	306	41918	4636.95	122.75	1.70	608887
		Pipe	40.64	3.99	47897328	48438214	258	35410	4636.98	122.75	1.70	610513

Table 7.7 Description: ASIC power, performance, and area (PPA) of different security levels of Rainbow signature generation algorithms. Gate Count is technology independent and is derived by computing $(Total\ Area - Memory\ Area)\ /\ (Nand2\ gate\ area)$. The Nand2 gate used to obtain the above results has an area of 2.1 μm^2 with GF 65 nm LPE process technology. Given a gate count and a memory size, the ASIC area can be estimated in any technology. FF Memory size in kilobytes gives the amount of memory implemented using Flip Flops instead of SRAM or SROM. Execution time can be derived by computing $(Clock Period \times Latency)$

Variant	Sec-level	Optimization	Power mW	Clock period ns	Area in μm^2 SRAM/ROM	Total area	Gate count $Kgate$	No of flip flops	Memory size in Kbytes SRAM	SROM	FF	Latency $Kcycles$
Ia_Classic	1	Base	7.36	4.04	1681938	2202195	248	37062	96.17	6.25	2.20	533
		Unroll	17.79	4.12	1681938	2369003	327	50341	96.17	6.25	2.20	494
		Pipe	7.38	4.43	1681938	2202155	248	37085	96.17	6.25	2.20	507
IIIc_Classic	3	Base	12.79	4.74	7391092	8163330	368	53591	509.33	6.25	3.45	2439
		Unroll	15.58	4.54	7391092	8286083	426	62549	509.33	6.25	3.45	2436
		Pipe	12.82	4.48	7391092	8163903	368	53640	509.33	6.25	3.45	2436
Vc_Classic	5	Base	18.57	4.07	16386585	17083846	332	48224	1166.17	6.25	3.42	5723
		Unroll	21.44	4.27	16386585	17225153	399	58945	1166.17	6.25	3.42	5716
		Pipe	18.88	4.51	16386585	17106910	343	50233	1166.17	6.25	3.42	5337

Table 7.8 **Description**: ASIC power, performance, and area (PPA) of different security levels of Rainbow signature verification algorithm. Gate Count is technology independent and is derived by computing $(Total\ Area - Memory\ Area)\ /\ (Nand2\ gate\ area)$. The Nand2 gate used to obtain the above results has an area of 2.1 μm^2 with GF 65 nm LPE process technology. Given a gate count and a memory size, the ASIC area can be estimated in any technology. FF Memory size in kilobytes gives the amount of memory implemented using Flip Flops instead of SRAM or SROM. Execution time can be derived by computing $(Clock\ Period \times Latency)$

Variant	Sec-level	Optimization	Power mW	Clock period ns	Area in μm^2		Gate count $K gate$	No of flip flops	Memory size in Kbytes			Latency $K cycles$
					SRAM/ROM	Total area			SRAM	SROM	FF	
`Ia_Classic`	1	Base	1.12	4.43	19046	195237	84	12463	0.38	0.00	0.83	244
		Unroll	1.75	4.68	19046	233785	102	14934	0.38	0.00	0.83	244
		Pipe	1.20	4.41	19046	199213	86	12769	0.38	0.00	0.83	156
`IIIc_Classic`	3	Base	1.24	4.48	20494	409458	185	27042	0.50	0.00	2.09	1299
		Unroll	1.38	4.46	20494	418103	189	27534	0.50	0.00	2.09	1298
		Pipe	1.25	4.51	20494	409438	186	27008	0.50	0.00	2.09	1298
`Vc_Classic`	5	Base	1.24	3.82	20494	334882	150	22605	0.50	0.00	1.77	3080
		Unroll	1.33	3.85	20494	343186	154	22884	0.50	0.00	1.77	3071
		Pipe	1.63	3.84	20494	351543	157	24012	0.50	0.00	1.77	1743

Table 7.9 Description: Area consumed by the design blocks used in Rainbow. Because of the issue when converting to HLS-ready C code, most of the functional block is made as part of top-level code. There are no smaller functions in the modified code. Hence, NA is mentioned against some of the blocks

	% of total area								
	Ia_Classic			IIIc_Classic			Vc_Classic		
Design blocks	Base	Unroll	Pipe	Base	Unroll	Pipe	Base	Unroll	Pipe
Key generation									
calculate_Q_from_F_r	5.3	4.3	5.3	3.4	2.7	3.4	3.3	2.5	3.3
prng_set	2.7	2.2	2.7	NA	NA	NA	NA	NA	NA
randombytes_with_sta	2.2	1.8	2.2	NA	NA	NA	NA	NA	NA
Signature generation									
prng_set	6.4	5.9	6.4	NA	NA	NA	NA	NA	NA
randombytes_with_sta	10.2	9.5	10.2	2.8	2.8	2.8	NA	NA	NA
p_hash	9.6	8.7	9.6	NA	NA	NA	NA	NA	NA
expand_hash	NA	NA	NA	1.8	1.8	1.8	NA	NA	NA
Signature verification									
p_hash_1	36.4	30.4	35.8	16.5	16.2	16.5	19.5	19.2	18.6
p_hash	36.0	30.1	35.3	33.8	33.1	33.8	19.5	19.0	18.4
batch_quad_trimat_ev	14.7	22.7	16.5	7.3	7.2	7.3	8.9	9.7	13.2

- For the three components (key generation, signature generation, and signature verification), loop unrolling requires more area compared to the other optimizations, and it requires less latency than the baseline implementation.
- For key generation, Table 7.3 shows that loop unrolling optimizations improve the latency by only $\sim$2%, increases LUTs by $\sim$6%, and increases FFs by $\sim$10%. Loop pipelining improves the latency by only 1% with little area overhead.
- Figure 7.1 shows design points with different optimizations. However, optimizations do not improve the latency relative to the baseline implementation.
- For signature generation, Table 7.4 shows that loop unrolling improves the latency by $\sim$6%, consume 1.5$\times$ more LUTs, and 2$\times$ more FFs.
- Signature generation does not use DSP modules. BRAMs are fixed for loop unrolling and loop pipelining since it is mainly used for faster data transfer.
- Table 7.4 shows that loop pipelining provides better optimization in terms of reduced latency and minimal area overhead.
- Figure 7.2 shows that loop pipelining drastically improves the latency for `Vc_Classic` (i.e., security level 5). Loop unrolling does not improve the latency, though it consumes a larger area to support parallelism.
- For signature verification, Table 7.5 shows that loop unrolling on `Ia_Classic` (i.e., security level 1) does not improve the latency, but it consumes 3$\times$ more LUTs, 3$\times$ more FFs, and uses one more BRAM.
- Signature verification does not use any DSP modules. BRAMs are fixed for loop unrolling and loop pipelining since these are mainly used for fast data transfer.

- Table 7.5 shows that loop pipelining is a better optimization, reduces latency and area overhead. For `Vc_Classic`, it reduces latency by ~50%.
- For ASIC implementation, the average clock period across the variant is 4.3 ns, which corresponds to 233 MHz. When compared to the 28 nm FPGA-based frequency numbers in Table 7.3, ASIC in 65 nm technology is almost 100% faster.
- Figures 7.4, 7.5, and 7.6 show that the higher security level has higher latency and power utilization.
- Rainbow ASIC DSE figure also indicates that optimization directives require more area and power. However, these optimizations do not improve latency for key generation and signature generation.
- With evaluation and exploration Rainbow hardware designs, we completed exploration of three multivariate PQC signature schemes. In next chapters, we will delve into hash-based signature schemes.

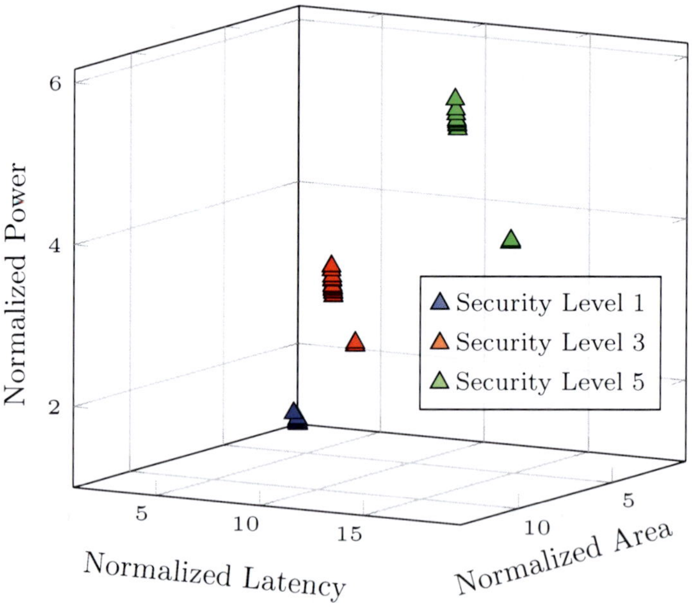

Fig. 7.4 ASIC design-space exploration of Rainbow key generation component when implemented using loop unrolling and pipelined optimizations, normalized to a baseline of Security level 1 key generation

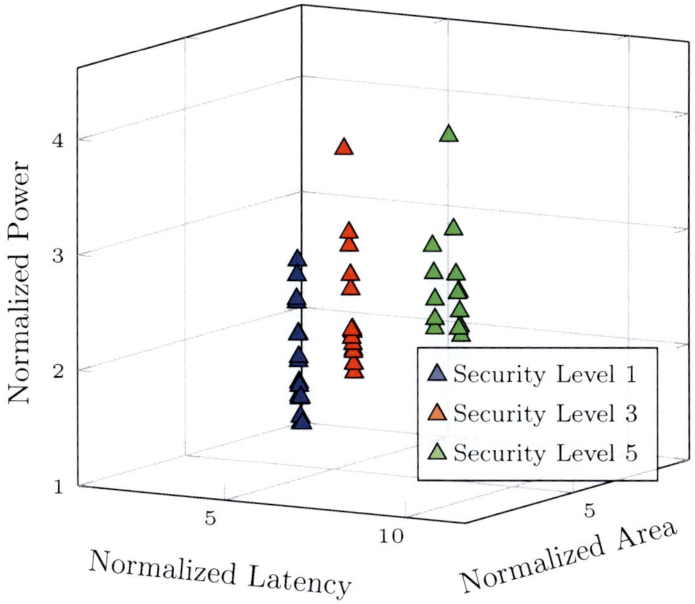

Fig. 7.5 ASIC design-space exploration of Rainbow signature generation component when implemented using loop unrolling and pipelined optimizations, normalized to a baseline of security level 1 signature generation

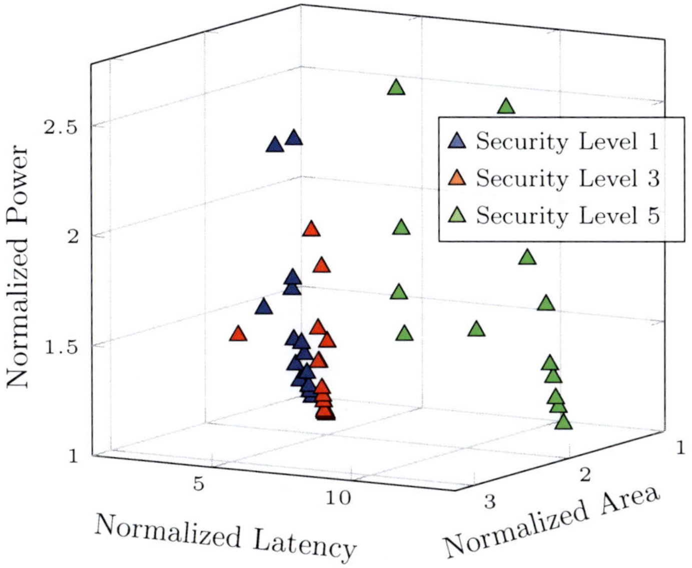

Fig. 7.6 ASIC design-space exploration of Rainbow signature verification component when implemented using loop unrolling and pipelined optimizations, normalized to a baseline of security level 1 signature verification

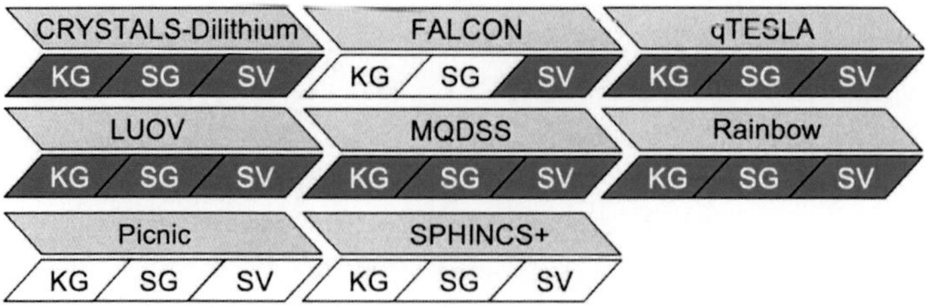

References

1. J. Ding, D. Schmidt, Rainbow a new multivariable polynomial signature scheme, in *International Conference on Applied Cryptography and Network Security* (Springer, New York, 2005), pp. 164–175
2. J. Ding, M.-S. Chen, A. Petzoldt, D. Schmidt, B.-Y. Yang, *Rainbow*. Submission to the NIST Post-Quantum Cryptography Standardization Project (2019). https://csrcnistgov/CSRC/media/Projects/Post-Quantum-Cryptography/documents/round-2/submissions/Rainbow-Round2.zip
3. A. Kipnis, J. Patarin, L. Goubin, Unbalanced oil and vinegar signature schemes, in *International Conference on the Theory and Applications of Cryptographic Techniques* (Springer, New York, 1999), pp. 206–222
4. K. Sakumoto, T. Shirai, H. Hiwatari, On provable security of UOV and hfe signature schemes against chosen-message attack, in *International Workshop on Post-Quantum Cryptography* (Springer, New York, 2011), pp. 68–82
5. A. Petzoldt, S. Bulygin, J. Buchmann, Cyclicrainbow–a multivariate signature scheme with a partially cyclic public key, in *International Conference on Cryptology in India* (Springer, New York, 2010), pp. 33–48

Chapter 8
Picnic

8.1 Algorithm Description

Picnic is a digital signature algorithm built on a zero-knowledge proof, and symmetric key primitives like hash functions and block ciphers, with well-understood post-quantum security [1, 2]. The security reductions are to hash functions and symmetric block ciphers. A Picnic signature is a noninteractive zero-knowledge proof of knowledge of the secret key. The plaintext being signed is incorporated (via hashing) into the challenges of the proof of knowledge in such a way that only the holder of the secret key can output the proof. The signature length depends on the multiplicative complexity of the encryption scheme and on the specific technique used to construct a zero-knowledge proof of knowledge (from the field of secure multi-party computation). Picnic has a small public key size, large signatures, and both signing and verification are slow. Key generation, on the other hand, is efficient.

In Picnic, the chosen block cipher is LowMC (a highly parameterizable block cipher with a low number of multiplications) [3, 4], as the resulting signature size is smaller than the alternatives. The proof of knowledge is either a specialized version of ZKBoo [5], called ZKB++ [6], or the proof by Katz et al. [7]. Picnic comes in nine different variants, covering three different security levels (see Tables 8.1, 8.2, and 8.3). For each of the three security levels, there are two signature algorithms that use the ZKB++ proof system, one based on the Fiat-Shamir (FS) transform and the other based on the Unruh (UR) transform. In addition, there are three parameter sets that use the proof system from [7] and are referred to as `Picnic2`. Picnic is an Existential Unforgeability under Chosen Message Attacks secure (EUF-CMA-secure) scheme in both the ROM and QROM (for the UR variant and recently also for the FS variants [8]).

Next, we present the key generation, signature generation, and signature verification algorithms.

D. Soni et al., *Hardware Architectures for Post-Quantum Digital Signature Schemes*, https://doi.org/10.1007/978-3-030-57682-0_8

Table 8.1 Security parameters, maximum signature size, public key size, and secret key size of Picnic NIST security level 1 variants

	picnic-L1-FS	picnic-L1-UR	picnic2-L1-FS
NIST security level	1	1	1
Parameter S	128	128	128
Parameter n	128	128	128
Parameter s	10	10	10
Parameter r	20	20	20
Parameter ℓ_H	256	256	256
Parameter T	219	219	343
Parameter u	–	–	27
Max. signature size (bytes)	34032	53961	13802
Public key size (bytes)	32	32	32
Private key size (bytes)	16	16	16

Table 8.2 Security parameters, maximum signature size, public key size, and secret key size of Picnic NIST security level 3 variants

	picnic-L3-FS	picnic-L3-UR	picnic2-L3-FS
NIST security level	3	3	3
Parameter S	192	192	192
Parameter n	192	192	192
Parameter s	10	10	10
Parameter r	30	30	30
Parameter ℓ_H	384	384	384
Parameter T	329	329	570
Parameter u	–	–	39
Max. signature size (bytes)	76772	121845	29750
Public key size (bytes)	48	48	48
Private key size (bytes)	24	24	24

Table 8.3 Security parameters, maximum signature size, public key size, and secret key size of Picnic NIST security level 5 variants

	picnic-L5-FS	picnic-L5-UR	picnic2-L5-FS
NIST security level	5	5	5
Parameter S	256	256	256
Parameter n	256	256	256
Parameter s	10	10	10
Parameter r	38	38	38
Parameter ℓ_H	512	512	512
Parameter T	438	438	803
Parameter u	-	-	50
Max. signature size (bytes)	132856	209506	54732
Public key size (bytes)	64	64	64
Private key size (bytes)	32	32	32

Algorithm 1 Picnic: key generation

Input : No Input Required
Output : sk = (data, public key) and pk = (param, plaintext, ciphertext)

procedure KEY GENERATION ▷ crypto_sign_keypair(sk, pk)
  Generate random secret key data. ▷ picnic_random_bytes
  Generate random public key plaintext. ▷ picnic_random_bytes
  Ciphertext ← LowMC(plaintext, data). ▷ LowMCEnc
  Copy public key to secret key.
  buffer (pk) ← public key. ▷ picnic_write_public_key
  buffer (sk) ← secret key. ▷ picnic_write_private_key
  return (sk = (data, public key), pk = (param, plaintext, ciphertext))
end procedure

Key Generation Algorithm 1 illustrates the steps in the key generation algorithm to output the public key and the secret key. The public key includes the plaintext and ciphertext, along with a parameter that indicates which variant of Picnic is being implemented. If the parameter value is 1, it implies that the algorithm performs the key exchange security level 1 Fiat-Shamir (FS) transform. If the parameter value is 2, it implies that the algorithm performs key exchange for security level 1 Unruh's transform (similar to Fiat-Shamir (FS) transform). The security level of Picnic increases with the parameter value.

Initially, the key generation algorithm generates public key plaintext and secret key data using a random number generator. Public key plaintext and secret key data size depend on the parameter (or security level). To verify the algorithm with KAT results, a pseudo-random number generator with AES is used instead of a random number generator. `LowMCEnc` accepts secret key data as the key and encrypts the plaintext to output a ciphertext. The secret key stores the LowMC key for the plaintext and ciphertext of the public key. Hence, breaking the secret key from the public key is equivalent to breaking the LowMC encryption. The algorithm makes a copy of the public key within the secret key. `picnic_write_public_key` function stores the $param$, $plaintext$, and $ciphertext$ serially in buffer (pk). `picnic_write_private_key` function stores the $param$, $data$, $plaintext$, and $ciphertext$ serially in buffer (sk). Finally, the algorithm produces the keypair: public key (pk) and secret key (sk).

Signature Generation Algorithm 2 describes the steps for signature generation, to create a signature using the message and secret key. The first step is to de-serialize the input secret key. If the secret key is not de-serialized correctly, the algorithm rejects the secret key and outputs an error message. The $ComputeSeed$ function uses SHAKE as a Key Derivation Function (KDF) with an input ($sk\ data$), message, ciphertext, and plaintext. It populates the 256-bit salt and seeds, which are used for the list of views ($V[0..numMPCrounds - 1][0..2]$) and the list of commitments ($C[0..numMPCrounds - 1][0..2]$). Here, $numMPCrounds$ is the number of parallel repetitions needed for soundness of the proof of knowledge. For security level 1 of Picnic, the number of parallel repetitions are 219. Each iteration

Algorithm 2 Picnic: signature generation

Input : Message m, Message Length $mlen$, and Secret Key sk = (data, public key)
Output : Signature sm = (signature_len, m, signature), Signature Length $smlen$

procedure SIGNATURE GENERATION ▷ crypto_sign($sm, smlen, m, mlen, sk$)
    De-serialize the private key. ▷ picnic_read_private_key
    Seed computation for parallel iteration. ▷ computeSeeds
    for `(k=0; k < numMPCRounds; k++)` **do**
        Create three random tapes. ▷ createRandomTape
        Compute three input shares.
        Run multi-party computation for three players. ▷ runMPC
        $C \leftarrow$ SHAKE(SHAKE(seed), view). ▷ Commit
    end for
    Compute the challenge. ▷ `H3`
    Packed the proof. ▷ prove
    Signature $\leftarrow$ (challengeBits, salt, proofs). ▷ serializeSignature
    $sm \leftarrow$ (signature length, message, signature)
end procedure

of the loop shown in Algorithm 2 can be computed separately. For each iteration and each player, $createRandomTape$ computes

$$rand[k][j] = SHAKE(SHAKE(seed[k][j])||salt||k||j||output\_length);$$

where k is the number of iterations and j the number of players who vary between 0 and 2. Signature generation computes inputshare of view, denoted by $x[0..2]$, each of length n bits: $x[0]$ = first n bits of $rand[0]$, $x[1]$ = first n bits of $rand[1]$, and $x[2] = sk \oplus x[0] \oplus x[1]$. The algorithm simulates a multi-party computation (MPC) protocol to calculate LowMC encryption and stores the data in the views of three players. Even if one of the players is corrupted, the algorithm remains secure. For each player, SHAKE computes a digest using the seed and the player's view. The `H3` function computes the challenge by hashing the $outputshares$, $commitments$, $publickey$, and $message$. Based on the challenge bit $e_k \in \{0, 1, 2\}$, the algorithm makes a copy of the seed and views for the two players to signature proofs. Signature generation serializes $challengebits$, $salt$, and $proofs$ in a single 8-bit buffer. The complete signature includes the signature length, message, and 8-bit signature buffer.

Signature Verification Algorithm 3 reports the signature verification steps, to verify the signature using the public key. The first step is to de-serialize the public key and store the parameter, plaintext, and ciphertext. Similarly, the algorithm de-serializes the input signature to signature length, message, challenge bits e, salt, and proofs. If the signature is not de-serialized correctly, the algorithm rejects this broken signature. If the signature de-serialization is correct, the algorithm proceeds to verify the signature. For each parallel iteration k, there is a different challenge bit e_k. Based on the challenge bit, seeds for two different players are provided. Using these seeds, random tape and input shares of views are computed.

Algorithm 3 Picnic: signature verification

Input : Signature sm = (signature_len, m, signature), Signature Length $smlen$, Public Key pk = (plaintext, ciphertext)
Output : "Accept"/"Reject", Message m, Message Length $mlen$

```
procedure SIGNATURE VERIFICATION          ▷ crypto_sign_open(m, mlen, sm, smlen, pk)
    De-serialize the public key.                            ▷ picnic_read_public_key
    if De-serialize the signature = FAIL then               ▷ deserializeSignature
        return "Reject"
    end if
    for (k=0; k < numMPCRounds; k++) do
        Re-construct the views and run MPC.                 ▷ verifyProof
        Update the commitments.                             ▷ Commit
    end for
    Compute the challenge.                                  ▷ H3
    if (computed_challengebits != received_challengebits) then
        return "Reject"
    end if
    m ← sm
end procedure
```

Signature verification simulates the MPC protocol to compute LowMC encryption. It is similar to signature generation with the key difference being the LowMC encrypt circuit is run on two players instead of three players. This change makes the LowMC marginally different. The algorithm makes two commitments based on the reconstructed views of two players within an iteration. The list of output shares are updated as follows:

$$output[k][e_k] = view[0].outputShare,$$

$$output[k][e_k + 1] = view[1].outputShare, \text{ and}$$

$$output[k][e_k + 2] = view[0].outputShare \oplus view[1].outputShare \oplus \text{ciphertext of}(pk).$$

Signature verification recomputes the challenge with input $outputs[0..2]$, $commits[0..k]$, $salt$, pk, and $message$. If the recomputed challenge for all the parallel iterations is equal to the input challenge from the signature, the input signature is valid; otherwise the signature is rejected.

Takeaway: The parameter signifies which version of Picnic is being implemented. It is used in key generation and stored by de-serializing the public key during signature verification.

8.2 Reference C Code ⟶ HLS-Ready C Code

Picnic requires significant changes to make it HLS-ready. With 132 changes in the code as shown in Table 8.4, Picnic requires the most complex modifications amongst all the signature schemes in NIST PQC Round-2. Many of the C code changes in Picnic are dependent on one another, which makes the conversion process challenging. To overcome this, we make the C code flexible by using conditional directives that allow one to go back to the original functionality and code easily. Table 8.4 classifies Picnic C code changes into five categories.

- Picnic uses `memcpy, memset, memmove,` and `AES` functions. Each library function is replaced with identical functionality. `memcpy` is the most commonly used library function.
- Picnic exploits dynamic memory allocation for message, signature, public key, secret key, random tape, views, commitments, challenge bits, and temporary variables. We replace dynamic memory allocation with static memory.
- Picnic uses typecasting operations as the same array is used by different functions. We replace typecasting by copying data into a variable of the target data type.
- We change file operations for the testbench.
- Picnic employs complex data structures and performs complex operations on large constant arrays. We change the C code to simplify data structures and efficiently access the large constant arrays.

Takeaway: The main challenge in converting Picnic C code to make it HLS-ready is the complex structures in the C code. The code is modified by simplifying such complex structures.

Table 8.4 Picnic: changes to reference C code

Class	Number of changes
Replace a library function	68
Remove dynamic memory allocation	16
Modify complex pointer operation	19
Change file operation	3
Modify of code structure	26
Total	132

8.3 FPGA-Specific Implementations and Optimizations

Amongst the nine variants of Picnic, we have implemented `picnic-L1-FS`, `picnic-L3-FS`, and `picnic-L5-FS`, which have distinct security levels. We implemented hardware designs for these three variants and successfully verified them using C/RTL co-simulation. For the signature generation components of `picnic-L3-FS` and `picnic-L5-FS`, C/RTL co-simulation runs for an indefinite time. To accurately estimate the area and latency requirements, we apply the following procedure.

We synthesize the complete design to obtain the area overhead. For latency requirements, we divide the signature generation design into three parts. The first part contains the `H3` function, the second part contains the `"prove"` function, and the third part contains the entire signature generation except for the `H3` function and the loop running the prove function (in sign_picnic1 function). For `picnic-L5-FS`, we reduce the number of MPC rounds to 1 for C/RTL co-simulation. We synthesize and verify all three parts, then sum up their latency to obtain the latency for the signature generation component of `picnic-L3-FS`. Based on the `picnic-L3-FS` signature generation latency value of the third part, we estimate the latency for the third part of the `picnic-L5-FS`. As a final step, we sum up the latency of the three parts for the signature generation of `picnic-L5-FS`.

To synthesize the design on an FPGA, Picnic requires complex changes. Listings 8.1 and 8.2 demonstrate examples of such changes in the code. Listing 8.1 shows a code snippet which returns one value from a linear matrix stored in structure matrices_t. This matrix is stored in a data structure, for which, memory can be allocated the size of the matrix, dynamically, whereas the size of the matrix is fixed with constant global variables. For security level 1, the size of the linear matrix is 128×4 for each of the 20 rounds. Although the matrix stores 8-bit elements, 32 bits at a time are read out at a time. This typecasting from 8 bits to 32 bits along with the flexibility of array access makes the C code incapable of synthesis. To overcome this synthesis hurdle, we declare three constant global matrices to access the data using an index as shown in Listing 8.2.

For Picnic variants, Tables 8.5, 8.6, and 8.7 show results of all the three components: key generation, signature generation, and signature verification, respectively. We optimized the area for the hash functions and optimized the performance for AES, `matrix_mul`, and `xor_array`. Based on these optimizations, loop pipelining improves the latency without significant area overhead. Tables 8.5, 8.6, and 8.7 show that the area of the baseline implementation is small enough to fit into an Artix-7 FPGA. Figures 8.1, 8.2, and 8.3 display the design-space exploration graphs which are normalized against the baseline implementation, which only has AES-related area optimizations.

```
typedef struct matrices {
    size_t nmatrices;
    size_t rows;
    size_t columns;
    const uint32_t* data;
} matrices_t;
const uint8_t linearMatrices_L1[40960] = {0x57, 0x19, ...};
const uint8_t roundConstants_L1[320] = {0x59, 0x04, ...};
const uint8_t keyMatrices_L1[43008] = {0x6B, 0xA7, ...};
const matrices_t LMatrix_L1={20,128,4,(uint32_t*)
    linearMatrices_L1};
const matrices_t KMatrix_L1={21,128,4,(uint32_t*)keyMatrices_L1};
const matrices_t RConstants_L1={20,1,4,(uint32_t*)
    roundConstants_L1};
// Functions to return individual matricies and round constants
#define ROW_SIZE(m) ((m).columns)
#define MAT_SIZE(m) ((m).rows*ROW_SIZE(m))
#define GET_MAT(m, r) (&(m).data[(r)*MAT_SIZE(m)])
/* Return the LowMC linear matrix for this round */
const uint32_t* LMatrix(uint32_t round, paramset_t* params) {
        return GET_MAT(LMatrix_L1, round);
}
```

Listing 8.1 Code: LowMC linear matrix for `picnic-L1-FS`

```
const uint32_t temp_matrix[43008/4] = {0xfd89a76b, ...};
const uint32_t temp_matrix2[40960/4] = {0x2c801957, ...};
const uint32_t temp_matrix3[320/4] = {0x950f0459, ...};
```

Listing 8.2 Code: synthesizable LowMC matrix access for `picnic-L1-FS`

Takeaway: With an increase in security level, although the hardware overhead does not change significantly, we observe that latency increases for all three components.

8.4 ASIC-Specific Implementations and Optimizations

In Table 8.8, the baseline implementation provides a better throughput for security level 1; however, for the other two variants, the pipelined version shows a better execution time. Power and area are higher for the loop unrolled implementation,

Table 8.5 Description: Performance, area, and security (PAS) trade-off of the Picnic key generation component for Artix-7 FPGA. Vivado-HLS directive loop unrolling and loop pipelining give various designs, which have different latency and area requirements, for three Picnic variants (`picnic-L1-FS`, `picnic-L3-FS`, and `picnic-L5-FS`)

Variant	Sec. level	Optimization	FF	LUT	BRAM	Clock DSP	Latency (ns)	*Kcycles*
`picnic-L1-FS`		Base	2137	11026	82	0	8.36	125
	1	Unroll	15793	41813	144	0	8.36	84
		Pipe	2139	11659	82	0	8.36	103
`picnic-L3-FS`		Base	2095	11179	274	0	8.36	354
	3	Unroll	15690	40909	337	0	8.36	241
		Pipe	2097	11812	274	0	8.36	287
`picnic-L5-FS`		Base	2118	11157	530	0	8.36	738
	5	Unroll	15549	43339	591	0	8.36	501
		Pipe	2113	11790	530	0	8.36	591

Table 8.6 Description: Performance, area, and security (PAS) trade-off of Picnic signature generation component for Artix-7 FPGA. Vivado-HLS directive loop unrolling and loop pipelining give various designs, which have different latency and area requirements, for three Picnic variants (`picnic-L1-FS`, `picnic-L3-FS`, and `picnic-L5-FS`)

Variant	Sec. level	Optimization	FF	LUT	BRAM	Clock DSP	Latency (ns)	*Kcycles*
`picnicl1fs`		Base	58932	142671	425	3	9.122	128521
	1	Unroll	61386	146945	497	3	9.122	111246
		Pipe	58940	142842	425	3	9.122	118096
`picnicl3fs`		Base	53576	152144	755	3	9.122	453951
	3	Unroll	54731	158480	757	3	9.122	372886
		Pipe	53673	152603	755	3	9.122	395816
`picnicl5fs`		Base	57846	157770	1175	3	9.122	902368
	5	Unroll	58227	159264	1177	3	9.122	841864
		Pipe	57850	157941	1175	3	9.122	859964

whereas the standard cell area is almost two times larger than other optimizations. All three variants have high SROM requirements as R constants, L Matrix, and K matrix store constant parameter values in SROM. The loop-unrolled version of `picnic-L5-FS` in particular requires an SROM size of more than one MBytes.

Similarly, Table 8.9 shows that the ROM requirement is higher for the loop unrolled implementation. The memory implemented with Flip Flops (FF column) is larger in size: there is a requirement for more than 250 memory space allocation with a size less than the minimum size, below which conversion to SRAM/SROM does not provide any area benefit. The pipelined optimization provides a better execution time; it also enables better latency and clock period, with respect to the other two implementations.

In Table 8.10, for security level 1, the loop-unrolled implementation has a better throughput, while the pipelined optimization gives a better execution time

Table 8.7 Description: Performance, area, and security (PAS) trade-off of Picnic signature verification component for Artix-7 FPGA. Vivado-HLS directive loop unrolling and loop pipelining give various designs, which have different latency and area requirements, for three Picnic variants (`picnic-L1-FS`, `picnic-L3-FS`, and `picnic-L5-FS`)

Variant	Sec. level	Optimization	FF	LUT	BRAM	Clock DSP	Latency (ns)	*Kcycles*
`picnic-L1-FS`	1	Base	57261	139129	404	2	9.122	87971
		Unroll	58728	142326	475	2	9.122	76283
		Pipe	57286	139627	404	2	9.122	81001
`picnic-L3-FS`	3	Base	44478	129324	686	2	9.122	307854
		Unroll	45644	135791	689	2	9.122	253404
		Pipe	44564	129882	686	2	9.122	269047
`picnic-L5-FS`	5	Base	48034	134018	1026	2	9.122	772961
		Unroll	51811	141699	2574	2	9.122	582449
		Pipe	49142	139097	1031	2	9.122	582135

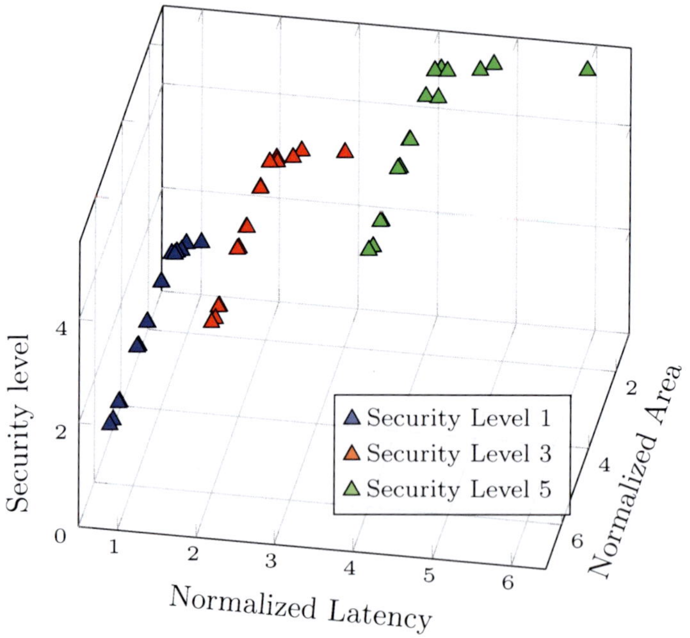

Fig. 8.1 FPGA design-space exploration of Picnic key generation component using loop unrolling and loop pipelining optimizations. Latency and area are normalized with respect to security level 1 baseline implementation (which contains no performance optimizations)

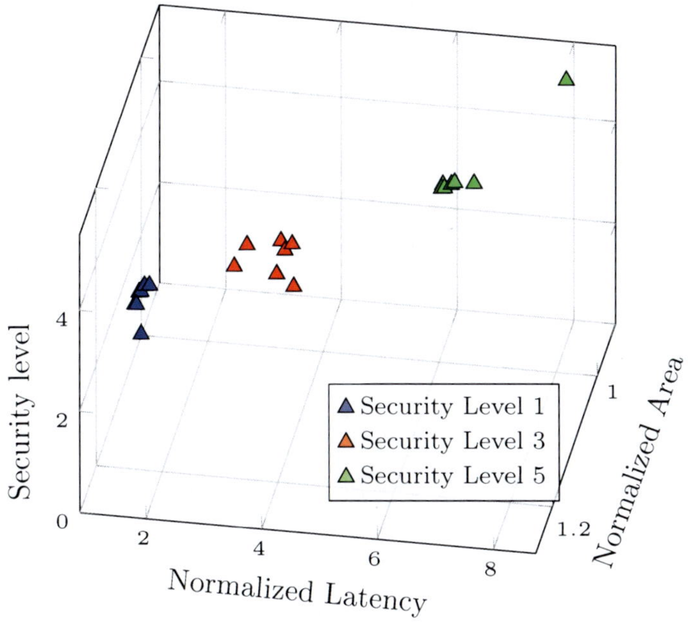

Fig. 8.2 FPGA design-space exploration of Picnic signature generation component using loop unrolling and loop pipelining optimizations. Latency and area are normalized with respect to security level 1 baseline implementation (which contains no performance optimizations)

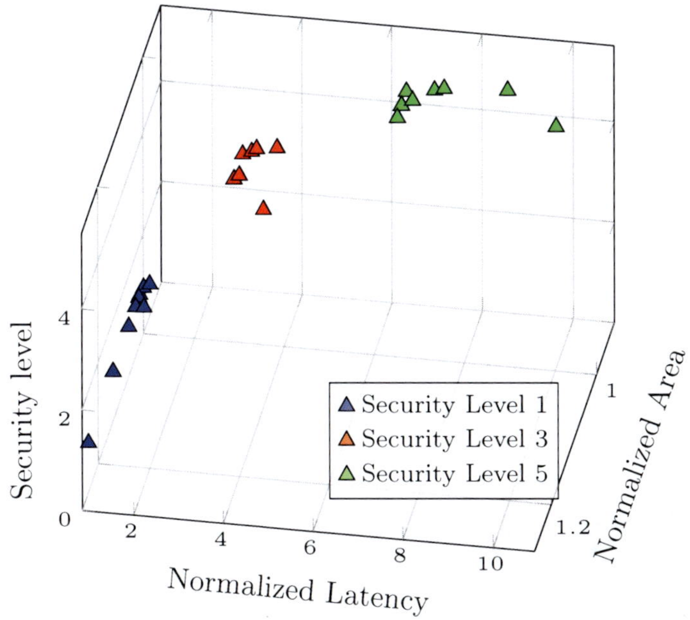

Fig. 8.3 FPGA design-space exploration of Picnic signature verification component using loop unrolling and loop pipelining optimizations. Latency and area are normalized with respect to security level 1 baseline implementation (which contains no performance optimizations)

Table 8.8 Description: ASIC power, performance, and area (PPA) of different security levels of Picnic key generation algorithm. Gate Count is technology independent and is derived by computing $(Total\ Area - Memory\ Area)\ /\ (Nand2\ gate\ area)$. The Nand2 gate used to obtain the above results has an area of 2.1 μm^2 with GF (Gate Forest Technology) 65 nm Low-Power Enhanced (LPE) process technology. Given a gate count and a memory size, the ASIC area can be estimated in any technology. FF Memory size in kilobytes gives the amount of memory implemented using Flip Flops instead of SRAM or SROM. Execution time can be derived by computing $(ClockPeriod \times Latency)$

Variant	Sec-level	Optimization	Power mW	Clock period ns	Area in μm^2 SRAM/ROM	Area in μm^2 Total area	Gate count $Kgate$	No of flip flops	Memory size in Kbytes SRAM	Memory size in Kbytes SROM	Memory size in Kbytes FF	Latency $Kcycles$
picnic-L1-FS	1	Base	1.83	3.63	250567	417656	79.57	15165	0.47	84.50	1.57	125
		Unroll	3.51	3.74	426920	763776	160.41	27798	0.47	167.00	1.54	84
		Pipe	1.75	4.26	227042	397074	80.97	15307	0.47	83.50	1.57	103
picnic-L3-FS	3	Base	2.12	3.69	564730	734831	81.00	14764	0.47	277.00	1.52	354
		Unroll	3.93	4.99	1055247	1390929	159.85	27303	0.47	552.00	1.49	241
		Pipe	2.05	4.11	541205	714060	82.31	14906	0.47	276.00	1.52	287
picnic-L5-FS	5	Base	2.22	4.84	1134082	1303919	80.87	14769	0.47	619.69	1.52	738
		Unroll	4.18	5.04	2215897	2533308	151.15	25421	0.47	1239.75	1.49	501
		Pipe	2.19	4.12	1110557	1284021	82.60	14904	0.47	618.69	1.52	591

Table 8.9 **Description**: ASIC power, performance, and area (PPA) of different security levels of Picnic signature generation algorithm. Gate Count is technology independent and is derived by computing $(Total\ Area - Memory\ Area) / (Nand2\ gate\ area)$. The Nand2 gate used to obtain the above results has an area of 2.1 μm^2 with GF 65 nm LPE process technology. Given a gate count and a memory size, the ASIC area can be estimated in any technology. FF Memory size in kilobytes gives the amount of memory implemented using Flip Flops instead of SRAM or SROM. Execution time can be derived by computing $(Clock Period \times Latency)$

Variant	Sec-level	Optimization	Power mW	Clock period ns	Area in μm^2		Gate count $K gate$	No of flip flops	Memory size in Kbytes			Latency $K cycles$
					SRAM/ROM	Total area			SRAM	SROM	FF	
picnic-L1-FS	1	Base	16.69	4.41	1154703	3659560	1192.79	204559	133.31	82.00	18.05	128521
		Unroll	17.09	4.28	1556344	4089672	1206.35	207192	133.31	246.0	18.05	111246
		Pipe	16.69	4.40	1154703	3659792	1192.90	204567	133.31	82.00	18.05	118096
picnic-L3-FS	3	Base	16.43	4.03	2551126	5239015	1279.95	223189	300.48	274.50	20.71	453951
		Unroll	17.40	4.99	3029881	5803343	1320.70	228010	300.48	549.00	21.14	372886
		Pipe	16.60	4.02	2551126	5240775	1280.78	223294	300.48	274.50	20.71	395816
picnic-L5-FS	5	Base	17.59	4.11	4558035	7302850	1307.05	227261	519.65	617.19	20.88	902368
		Unroll	17.99	4.99	5606141	8361256	1311.96	227789	519.65	1234.38	20.88	841864
		Pipe	17.60	4.06	4558035	7302491	1306.88	227265	519.65	617.19	20.88	859964

Table 8.10 **Description**: ASIC power, performance, and area (PPA) of different security levels of Picnic signature verification algorithms. Gate Count is technology independent and is derived by computing $(Total\ Area - Memory\ Area)\ /\ (Nand2\ gate\ area)$. The Nand2 gate used to obtain the above results has an area of 2.1 μm^2 with GF 65 nm LPE process technology. Given a gate count and a memory size, the ASIC area can be estimated in any technology. FF Memory size in kilobytes gives the amount of memory implemented using Flip Flops instead of SRAM or SROM. Execution time can be derived by computing $(ClockPeriod \times Latency)$

Variant	Sec-level	Optimization	Power mW	Clock period ns	Area in μm^2		Gate count $K gate$	No of flip flops	Memory size in Kbytes			Latency $K cycles$
					SRAM/ROM	Total area			SRAM	SROM	FF	
picnic-L1-FS	1	Base	16.09	4.61	962169	3361718	1142.64	195445	113.50	82.00	17.11	87971
		Unroll	16.58	4.40	1472170	3808311	1112.45	189181	114.48	246.00	16.14	76283
		Pipe	16.07	4.55	962169	3363919	1143.69	195446	113.50	82.00	17.11	81001
picnic-L3-FS	3	Base	13.83	4.61	2150057	4374143	1059.09	184479	256.22	274.50	17.08	307854
		Unroll	15.10	5.04	2794801	5034164	1066.36	182861	257.20	549.00	16.74	253404
		Pipe	13.86	4.49	2150057	4 375426	1059.70	184566	256.22	274.50	17.08	269047
picnic-L5-FS	5	Base	14.54	4.41	3821214	6092052	1081.35	188158	444.56	617.19	17.26	772961
		Unroll	16.41	5.03	9450042	11688631	1065.99	185953	445.53	3700.75	16.48	582449
		Pipe	14.56	4.44	3821214	6097446	1083.92	188198	444.56	617.19	17.26	582135

Table 8.11 Description: Area consumed by the design blocks used in Picnics

	% of total area								
	picnic-L1-FS			picnic-L3-FS			picnic-L5-FS		
Design blocks	Base	Unroll	Pipe	Base	Unroll	Pipe	Base	Unroll	Pipe
Key generation									
LowMCEnc	65.3	72.6	68.8	80.6	85.3	82.9	89.1	92.0	90.5
randombytes	28.4	23.6	24.3	15.7	12.5	13.0	8.8	6.9	7.2
Signature generation									
H3_2	17.3	15.6	17.4	11.2	10.2	11.2	8.2	7.2	8.2
mpc_LowMC_2	6.4	14.2	6.4	10.5	19.0	10.5	15.4	26.2	15.4
Commit_2	14.9	13.4	14.9	9.5	8.7	9.5	7.0	6.1	7.0
createRandomTape	NA	NA	NA	9.3	8.4	9.3	6.8	5.9	6.8
createRandomTape_1	NA	NA	NA	9.3	8.4	9.3	6.8	5.9	6.8
computeSeeds_2	12.8	11.5	12.8	8.6	7.9	8.6	6.2	5.4	6.2
Signature verification									
H3_2	18.9	16.7	18.9	13.4	11.7	13.5	9.8	5.1	9.9
Commit_23	16.2	14.3	16.2	11.4	9.9	11.4	8.4	4.3	8.4
createRandomTape13	15.8	13.8	15.8	11.2	9.7	11.2	8.1	4.2	8.1
mpc_LowMC_verify_2	6.9	15.1	6.9	12.5	21.8	12.5	18.4	54.5	18.4

for the other two variants. ROM requirements are higher: loop unrolled version of picnic-L5-FS necessitates more than three megabytes of SROM. Moreover, memory implemented with Flip Flops (FF column) is larger in size: there is a requirement for more than 200 memory space allocation with a size less than the minimum size, below which conversion to SRAM/SROM does not provide any area benefit.

Table 8.11 reports high area requirements for internal Picnic modules for key generation, signature generation, and signature verification, respectively. Based on these reports, key design blocks in Picnic are LowMCEnc, randombytes, H3_2, mpc_LowMC_2, commit_2, createRandomTape, computeSeeds_2, and mpc_LowMC_verify_2.

Takeaway: For security level 1, the baseline version provides the best throughput, but for security levels 3 and 5, the pipelined implementations have shorter execution times.

8.5 Key Takeaways

- Picnic FPGA components have a minimum clock period of 8.36 ns, yielding a maximum frequency of $\sim$120 MHz.
- In relation to the dimension of area overhead and execution latency, the order is: key generation < signature verification < signature generation.
- In terms of latency, the order is: security level 1 < security level 3 < security level 5.
- Loop unrolling has the least latency with an increase in area overhead. Loop pipelining reduces the latency with minimum area overhead.
- For key generation, Table 8.5 shows that loop unrolling improves the latency by $\sim$35%, increases LUTs 4$\times$, and increases FFs 7$\times$, when compared to the baseline implementation. Loop pipelining improves latency by $\sim$20%, with a negligible area overhead compared to the baseline.
- Table 8.5 shows that the key generation component requires no memory.
- Figure 8.1 reveals several design points with different optimizations.
- For signature generation, Table 8.6 shows that loop unrolling improves latency by $\sim$15%, increases LUTs by $\sim$3%, and increases FFs by $\sim$3%.
- Table 8.6 shows that latency increases by $\sim$2–3$\times$ with higher security levels. Area overhead increases by $\sim$7–8% with higher security levels.
- Figure 8.2 shows that loop pipelining and loop unrolling slightly reduce the latency for all security levels. The outlier design point for each security level shows the design without any area or performance optimizations.
- For signature verification, Table 8.7 shows that latency increases by $\sim$3$\times$ with higher security levels. Considering LUTs and FFs (excluding DSPs and BRAMs), security level 3 < security level 5 < security level 1.
- Table 8.7 shows that loop pipelining offers the best optimization results for security level 5, and loop unrolling provides better optimization results for security level 1 and security level 3.
- Figure 8.3 shows that loop pipelining and loop unrolling reduce the latency. The algorithm follows the expected area-speed trade-off curve. The outlier design point for each security level shows the design with no area or performance optimizations.
- The number of flip flops is more in ASIC than the FPGA equivalent. This is because all the memory is implemented using BRAM in FPGA. But in ASIC, the memory is converted to SRAM only if the conversion gives area gains; thus, smaller memory is not implemented as SRAM, but with flip flops.
- For ASIC implementation, the SRAM requirement is less and remains the same for all the security levels for the key generation algorithm compared to other key exchange components, where SRAM size is large and increases with higher security levels.
- Figures 8.4, 8.5, and 8.6 show that the higher the security level, the higher the latency.

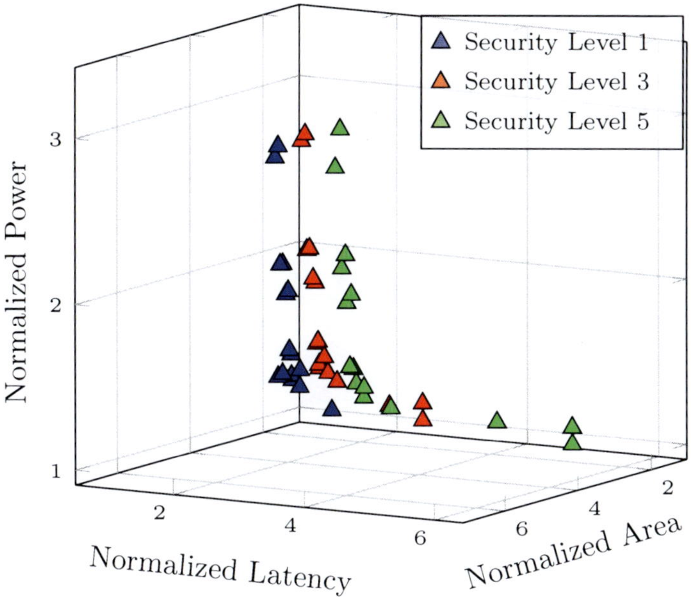

Fig. 8.4 ASIC design-space exploration of Picnic key generation component when implemented using loop unrolling and pipelined optimizations, normalized to a baseline of security level 1 key generation

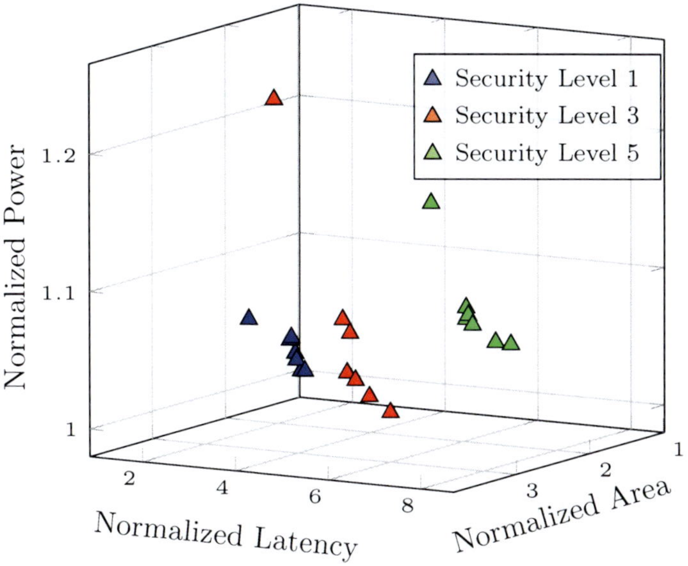

Fig. 8.5 ASIC design-space exploration of Picnic signature generation component when implemented using loop unrolling and pipelined optimizations, normalized to a baseline of Security level 1

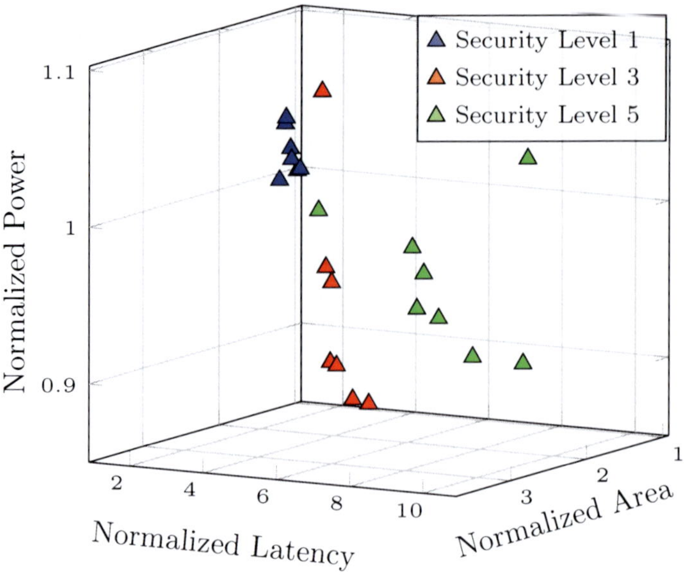

Fig. 8.6 ASIC design-space exploration of Picnic signature verification component when implemented using loop unrolling and pipelined optimizations, normalized to a baseline of Security level 1 signature verification

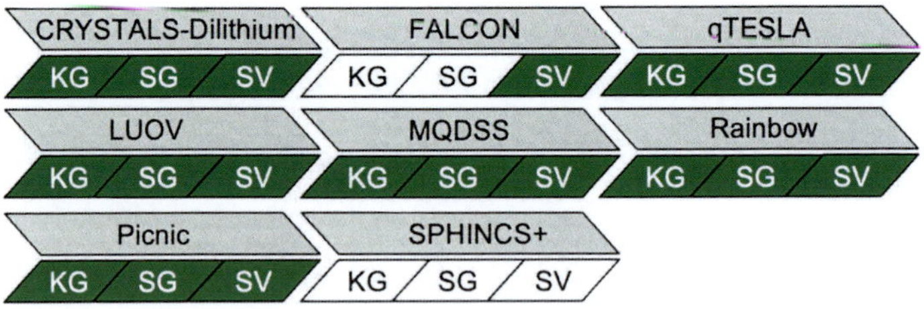

- In this chapter, we evaluated and explored the hardware design of hash-based Picnic signature scheme.

References

1. G. Zaverucha, M. Chase, D. Derler, S. Goldfeder, C. Orlandi, S. Ramacher, C. Rechberger, D. Slamanig, J. Katz, X. Wang, V. Kolesnikov, D. Kales, *Picnic*. Submission to the NIST Post-Quantum Cryptography Standardization Project (2019). https://csrc.nist.gov/CSRC/media/Projects/Post-Quantum-Cryptography/documents/round2/submissions/Picnic-Round2.zip
2. M. Chase, D. Derler, S. Goldfeder, C. Orlandi, S. Ramacher, C. Rechberger, D. Slamanig, G. Zaverucha, The picnic signature algorithm specification. March 2019 [Online]. Available: https://github.com/microsoft/Picnic/blob/master/spec/specv2.1.pdf

3. M.R. Albrecht, C. Rechberger, T. Schneider, T. Tiessen, M. Zohner, Ciphers for MPC and FHE, in *Annual International Conference on the Theory and Applications of Cryptographic Techniques* (Springer, New York, 2015), pp. 430–454
4. I. Dinur, D. Kales, A. Promitzer, S. Ramacher, C. Rechberger, Linear equivalence of block ciphers with partial non-linear layers: application to LowMC, in *Annual International Conference on the Theory and Applications of Cryptographic Techniques* (Springer, New York, 2019), pp. 343–372
5. I. Giacomelli, J. Madsen, C. Orlandi, ZKBoo: faster zero-knowledge for Boolean circuits, in *25th USENIX Security Symposium (USENIX Security 16)* (2016), pp. 1069–1083
6. M. Chase, D. Derler, S. Goldfeder, C. Orlandi, S. Ramacher, C. Rechberger, D. Slamanig, G. Zaverucha, Post-quantum zero-knowledge and signatures from symmetric-key primitives," in *Proceedings of the 2017 ACM SIGSAC Conference on Computer and Communications Security* (ACM, New York, 2017), pp. 1825–1842
7. J. Katz, V. Kolesnikov, X. Wang, Improved non-interactive zero knowledge with applications to post-quantum signatures, in *Proceedings of the 2018 ACM SIGSAC Conference on Computer and Communications Security* (ACM, New York, 2018), pp. 525–537
8. J. Don, S. Fehr, C. Majenz, The measure-and-reprogram technique 2.0: multi-round Fiat-Shamir and more (2020). Preprint. arXiv:2003.05207

Chapter 9
SPHINCS$^+$

9.1 Algorithm Description

SPHINCS$^+$ [1, 2] is a stateless hash-based signature scheme that was submitted to the NIST competition. At a high level, SPHINCS$^+$ works much like SPHINCS [3]. The design rationale follows the original SPHINCS and applies several modifications based on the analyses of recent research: (a) multi-target attack protection [4]; (b) tree-less WOTS$^+$ public key compression using a single tweakable hash function call; and (c) Forest of Random Sets (FORS), an improvement of HORST [5] which in turn is a variant of HORS [3]. SPHINCS$^+$ has the most conservative design for a post-quantum hash-based signature scheme and therefore, it features large signature sizes and slow operations (slow speed)—the main disadvantages of this scheme.

SPHINCS$^+$ is an Existential Unforgeability under Adaptive Chosen Message Attacks (EUF-CMA) secure scheme, and its security is solely based on standard properties of the used function families and the assumption that the Pseudo-Random Function (PRF) used within the instantiations of the tweakable hash functions can be modeled as a random oracle. The security claim is based on other works [2, 6]. In addition, a collision attack against the hash function does not suffice to break the security. Tables 9.1, 9.2, and 9.3 summarize the different SPHINCS$^+$ parameter sets spanning three security levels, where *-f* and *-s* stand for *fast* and *small* variants, respectively.

SPHINCS$^+$ key generation, signature generation, and signature verification components are built on WOTS$^+$ and FORS schemes. The `Treehash` function is the basic building block of WOTS$^+$ and FORS. `Treehash` function is used to generate the authentication path and root node of the sub-tree. As shown in Algorithm 1, it computes the internal node of its Merkle tree using a secret seed (SK_SEED), a public seed (PUB_SEED), an index of the leaf ($leaf\_idx$), an index offset for FORS (idx_offset), the target node height ($tree\_height$), and an address of the tree ($tree\_addr$). Using the height of the tree node, the loop iterates $2^{tree\_heights}$ times. Each loop generates a leaf node, increases the offset, and sets the

D. Soni et al., *Hardware Architectures for Post-Quantum Digital Signature Schemes*, https://doi.org/10.1007/978-3-030-57682-0_9

Table 9.1 Security parameters, signature size, public key size, and secret key size of SPHINCS$^+$ NIST security level 1 variants

	SPHINCS$^+$-128f	SPHINCS$^+$-128s
NIST security level	1	1
Parameter n	16	16
Parameter h	60	64
Parameter d	20	8
Parameter $\log t$	9	15
Parameter k	30	10
Parameter w	16	16
Signature size (bytes)	16,976	8080
Public key size (bytes)	64	64
Private key size (bytes)	32	32

Table 9.2 Security parameters, signature size, public key size, and secret key size of SPHINCS$^+$ NIST security level 3 variants

	SPHINCS$^+$-192f	SPHINCS$^+$-192s
NIST security level	3	3
Parameter n	24	24
Parameter h	66	64
Parameter d	22	8
Parameter $\log t$	8	16
Parameter k	33	14
Parameter w	16	16
Signature size (bytes)	35,664	17,064
Public key size (bytes)	96	96
Private key size (bytes)	48	48

Table 9.3 Security parameters, signature size, public key size, and secret key size of SPHINCS$^+$ NIST security level 5 variants

	SPHINCS$^+$-256f	SPHINCS$^+$-256s
NIST security level	5	5
Parameter n	32	32
Parameter h	68	64
Parameter d	17	8
Parameter $\log t$	10	14
Parameter k	20	22
Parameter w	16	16
Signature size (bytes)	49,216	29,792
Public key size (bytes)	128	128
Private key size (bytes)	64	64

Algorithm 1 SPHINCS$^+$: Treehash

Input : SK_SEED, PUB_SEED, leaf_idx, idx_offset, tree_height, tree_addr
Output : SK = (root, auth_path)

```
1: procedure TREEHASH   ▷ treehash(root, auth_path, SK_SEED, PUB_SEED, leaf_idx, idx_offset, tree_height, tree_addr)
2:     for (idx=0; idx < 2^tree_height; idx++) do
3:         Add the leaf node to the root   ▷ wots_gen_leaf, fors_gen_leaf
4:         offset ← offset + 1; heights[offset -1] ← 0;
5:         Copy the root node as authenticating path if (leaf_idx xor 1 == idx)   ▷ memcpy
6:         while offset >= 2 & heights[offset-2] == heights[offset-1] do
7:             Set the address of the new node   ▷ set_tree_heights, set_tree_index
8:             Hash the top node   ▷ thash
9:             offset ← offset-1; heights[offset-1] ← heights[offset-1] + 1;
10:            Store for authenticating path if required   ▷ memcpy
11:        end while
12:    end for
13: end procedure
```

height of the leaf node to 0. If the leaf node is required for path authentication, the value is stored. Algorithm 1: line-6, runs a while loop to generate the nodes for the tree if two (2) leaves are added to the tree. Each pair of two leaves requires a parent node in a binary tree. We need to set an address to create the node. We generate the hash of the top-node of the stack by taking two nodes as input. The height of the new node is higher than the leaf node. If the new node is in the route for authentication, it gets stored. Repeating this node generation process helps to reach only one topmost node.

Algorithm 2 SPHINCS$^+$: key generation

Input : No Input Required
Output : SK = (SK_SEED, SK_PRF, PUB_SEED, root) and PK = (PUB_SEED, root)

```
procedure KEY GENERATION   ▷ crypto_sign_keypair(sk, pk)
    Generation of SK_SEED, SK_PRF, and PUB_SEED using AES   ▷ randombytes
    Set layer initialize   ▷ set_layer_addr, set_type
    Hash function initialization.   ▷ initialize_hash_function
    Compute the root node of the top layer of tree.   ▷ treehash
    return (SK, PK)
end procedure
```

Key Generation Creates a public key (PK) and a secret key (SK) as shown in Algorithm 2. The public key contains an n-byte public seed value and the root of the tree of the top layer. The secret key contains a seed to generate WOTS$^+$ and FORS private key elements, a pseudo-random function (PRF) key, and a public key. AES generates the random number for SK_SEED, SK_PRF, and PUB_SEED. Based on the SK_SEED and PUB_SEED, we need to create the sub-tree. First, we set the tree layer. We initialize the hash function based on the public seed and the secret seed. Finally, we compute the root node of the sub-tree as mentioned in Algorithm 1.

Algorithm 3 SPHINCS$^+$: signature generation

Input : Message m, Message Length $mlen$, and Secret Key sk = (SK_SSED, SK_PRF, PUB_SEED, root)
Output : Signature sm = (sign, m), Signature Length $smlen$

procedure SIGNATURE GENERATION ▷ crypto_sign($sm, smlen, m, mlen, sk$)
    Initialize the hash function with PUB_SEED and SK_SEED. ▷ initialize_hash_function
    Generate a random seed number. ▷ randombytes
    Compute the digest from random number, SK_PRF, and message. ▷ gen_message_random
    Compute message hash, tree and leaf index. ▷ hash_message
    Setup for FORS$^+$ and WOTS. ▷ set_type, set_tree_addr, set_keypair_addr
    Sign the message hash using FORS$^+$ ▷ fors_sign
    for `(i=0; i < SPX_D; i++)` **do**
        Setup the WOTS layer. ▷ set_layer_addr, set_tree_addr, copy_subtree_addr, set_keypair_addr
        Compute WOTS signature. ▷ wots_sign
        Compute hash root node tree and authentication path (part of signature). ▷ treehash
    end for
    Concatenate the message with sign.
    return (signature)
end procedure

Signature Generation Algorithm 3 shows signature generation component, which generates the digital signature using the secret key. First, we initialize the hash function; `SHA`, `HARAKA`, and `SHAKE` are different hash functions used for SPHINCS$^+$ variants. We then generate the random number which is used to generate random hash values for the message along with the message and SK_PRF. Based on this random hash value, the message digest ($mhash$), leaf index, and tree values are generated. The generated FORS$^+$ public key and signature are stored in the SPHINCS$^+$ signature. Now, SPX_D sub-trees and layers need to be created; SPX_D values vary with each variant. For each layer, the WOTS tree parameter is set up and we generate the WOTS signature. Using this signature, the algorithm computes the root and the authentication paths for the input WOTS leaf. The index leaf and tree are updated for the next layer or iteration. When the calculation of the authentication path for all sub-trees is over, the signature is considered complete.

Signature Verification Verifies the signature using the public key as shown in Algorithm 4. It takes PUB_SEED from the public key to initialize the hash function. If the signature length is not SPX_BYTES (size of message hash + FORS$^+$ signature size + all WOTS sub-tree signature size), the signature is not valid and hence is rejected. Otherwise, the WOTS tree is setup. The message digest, tree, and leaf index are generated using the signature random number, public seed, and message from the signature. Using this message digest, public seed, and

Algorithm 4 SPHINCS$^+$: signature verification

Input : Signature sm = (sign, m), Signature Length $smlen$, Public Key pk = (PUB_SEED, root)
Output : "Accept"/"Reject", Message m, Message Length $mlen$

```
procedure SIGNATURE VERIFICATION    ▷ crypto_sign_open(m, mlen, sm, smlen, pk)
    if smlen != SPX_BYTES then return "Reject"
    end if
    Initialize the hash function with PUB_SEED.    ▷ initialize_hash_function
    Setup WOTS tree.    ▷ set_type, set_tree_addr, set_keypair_addr, copy_keypair_addr
    Hash the message using random number, public seed and message.    ▷ hash_message
    Compute the FORS+ root and public key.    ▷ fors_pk_from_sig
    for (i=0; i < SPX_D; i++) do
        Setup the WOTS layer.    ▷ set_layer_addr, set_tree_addr, copy_subtree_addr, set_keypair_addr
        Compute WOTS public key.    ▷ wots_pk_from_sig
        Compute leaf node hash of the WOTS.    ▷ thash
        Compute the root node of the sub-tree.    ▷ compute_root
        Update values for next layer/loop.
    end for
    if root != (root from PK) then return "Reject"
    else"Accept"
        Copy message from sign.    ▷ memcpy
        return (Message)
    end if
end procedure
```

signature, the FORS public key is computed. There are SPX_D sub-trees. For each WOTS sub-tree, the root value is computed. The sub-tree layer is set up and we compute WOTS' public key. Using the public key, the hash value of the leaf node is computed. The leaf node and the authentication path from the key are used to compute the root node of the sub-tree. The value of the next sub-tree is then updated. If all the sub-trees are computed, the root node of the final sub-tree should match with the public key root value for successful signature verification. If it does not match, it serves as an indicator that the signature creator (signer) does not have the secret key and the signature is consequently updated.

Takeaway Once all sub-trees are computed, the root node of the final sub-tree is matched with the public key root value. If there is a match, signature verification is successful.

9.2 Reference C Code ⟶ HLS-Ready C Code

SPHINCS$^+$ uses a complex C design to generate the FORS and WOTS tree. We make several changes in the C code to make it capable of synthesis as reported in Table 9.4.

- SPHINCS$^+$ uses many `memcpy`, `memset`, and `memmove` functions. `memset` changes a specified value into a constant and `memmove` copies data from one array to another. Such library functions are replaced with identical functionality for the purposes of this exercise.
- The memory allocation in the function depends on the input variable of the function. We replace this dynamic memory allocation with static memory allocation to make it independent of any specific input.
- SPHINCS$^+$ uses a functional pointer to generate leaves for WOTS and FORS tree. We replace the function pointers with separate functions. We removed the pointers, which are not used for any operations, in `prf_addr` and `thash`.
- We change the file structure and dynamic memory allocation in the testbench.
- We modify the C code to accommodate global variables in the hardware design, to replace function pointers, and to replace functionality of the smaller modules such as `set_type`.

Takeaway SPHINCS$^+$ uses memory operations like `memcpy`, `memset`, and `memmove` extensively. These are replaced with identical functions which replicate the memory operations, in lieu of the library functions. Dynamic memory allocation is replaced by static allocation.

Table 9.4 SPHINCS$^+$: changes to reference C code

Class	Number of changes
Replacing library function	52
Removing dynamic memory allocation	6
Modifying complex pointer operation	2
Changing file operation	3
Modification of code structure	19
Total	82

9.3 FPGA-Specific Implementations and Optimizations

Among the different variations of SPHINCS$^+$, we have implemented sphincs-haraka-128f-simple, sphincs-haraka-192f-simple, and sphincs-haraka- 256f-simple. Each of these has different security strengths. For these three algorithms, we have implemented all the three components: key generation, signature generation, and signature verification. We have successfully verified each design point of key generation and signature verification modules using C/RTL co-simulation. Since signature generation takes a considerable amount of time for verification, all the design points could not be verified. Hence, we verified only baseline optimization of signature generation with Known Answer Tests (KATs). For other optimizations, we alter the C code to compute only one FORS tree and one WOTS tree. HLS is executed on this modified C code and the design is verified with the process mentioned in Fig. 1.1. Once a correct result is obtained for latency and area utilization, we extrapolate latency results for other optimizations.

```
void aesenc(unsigned char *s, const unsigned char *rk)
{
    unsigned char i, t, u, v[4][4];
    aesenc_label0:for (i = 0; i < 16; ++i) {
        v[((i / 4) + 4 - (i%4) ) % 4][i % 4] = sbox[s[i]];
    }
    aesenc_label1:for (i = 0; i < 4; ++i) {
        t = v[i][0];
        u = v[i][0] ^ v[i][1] ^ v[i][2] ^ v[i][3];
        v[i][0] ^= u ^ XT(v[i][0] ^ v[i][1]);
        v[i][1] ^= u ^ XT(v[i][1] ^ v[i][2]);
        v[i][2] ^= u ^ XT(v[i][2] ^ v[i][3]);
        v[i][3] ^= u ^ XT(v[i][3] ^ t);
    }
    aesenc_label2:for (i = 0; i < 16; ++i) {
        s[i] = v[i / 4][i % 4] ^ rk[i];
    }
}
```

Listing 9.1 AES encryption

SPHINCS$^+$ or XMSS schemes rely on the security of hash functions, but the latter creates performance bottlenecks for these schemes (Table 9.5). In order to address this bottleneck, the haraka hash function using AES is used, which expedites the implementation [7] as seen in Figs. 9.1, 9.2, and 9.3.

Listing 9.2 shows haraka512, which is a haraka hash function that is employed by all the haraka variants. It takes 512 bits of input (64 bytes) and produces 512 bits of hash output (64 bytes). It performs a permutation (using haraka512_perm)

Table 9.5 Description: Performance, area, and security (PAS) trade-off of SPHINCS$^+$ key generation component for Artix-7 FPGA. Vivado-HLS directive loop unrolling and loop pipelining give various designs, which have different latency and area requirements, for three SPHINCS$^+$ variants (`sphincs-haraka-128f-simple`, `sphincs-haraka-192f-simple`, and `sphincs-haraka-256f-simple`)

Variant	Sec. level	Optimization	FF	LUT	BRAM	DSP	Clock (ns)	Latency *Kcycles*
`sphincs-haraka-128f-simple`	1	Base	6812	28,590	37	0	6.62	34,364
		Unroll	9077	37,590	57	0	7.727	18,507
		Pipe	6812	28,933	38	0	6.62	25,969
`sphincs-haraka-192f-simple`	3	Base	6882	28,791	37	0	6.62	50,795
		Unroll	9147	37,791	57	0	7.727	27,453
		Pipe	6882	29,134	38	0	6.62	38,438
`sphincs-haraka-256f-simple`	5	Base	6432	28,147	44	0	6.62	135,476
		Unroll	8697	37,165	64	0	7.727	73,455
		Pipe	6432	28,490	45	0	6.62	102,646

```
void haraka512(unsigned char out[64], const unsigned char in[64])
{
    int i, loop;
    unsigned char buf[64];

    haraka512_perm(buf, in);
    /* Feed-forward */
    haraka512_label3:for (i = 0; i < 64; i++) {
        buf[i] = buf[i] ^ in[i];
    }
    /* Truncated */
    haraka512_label4:for(loop=0;loop<8;loop++)
      out[loop] = buf[8+loop];
    haraka512_label5:for(loop=0;loop<8;loop++)
      out[loop+8] = buf[24+loop];
    haraka512_label6:for(loop=0;loop<8;loop++)
      out[loop+16] = buf[32+loop];
    haraka512_label7:for(loop=0;loop<8;loop++)
      out[loop+24] = buf[48+loop];
//memcpy(out,      buf + 8, 8); memcpy(out + 8,  buf + 24, 8);
//memcpy(out + 16, buf + 32, 8); memcpy(out + 24, buf + 48, 8);
}
```

Listing 9.2 Haraka hash function of SPHINCS$^+$

and computes XOR of `haraka512` input and `haraka512_perm` output, to generate an output hash. `haraka512_perm` uses four parallel AES encryption operations for the permutation, as AES can only operate on 16 bytes at a time. AES can either take a fixed round key value, or a different round key value, that was tweaked during hash initialization. We have performed loop unrolling and loop pipelining optimizations on `aesenc` and `haraka512` function. Listings 9.1 and 9.2 present the loops on which we have used HLS directives for optimization. These directives have resulted in $\sim$50% latency improvement in the designs, as reported in Tables 9.5, 9.6, and Table 9.7. Figures 9.1, 9.2, and 9.3 report the area overhead and latency of different designs created by loop unrolling and loop pipelining on `aesenc` and `haraka512`. We have partially optimized the hash function(s) of SPHINCS$^+$. To explore devices with the least hardware overhead or latency, we explore and improve all the functions/loops of SPHINCS$^+$, including the hash functions (Tables 9.6 and 9.7).

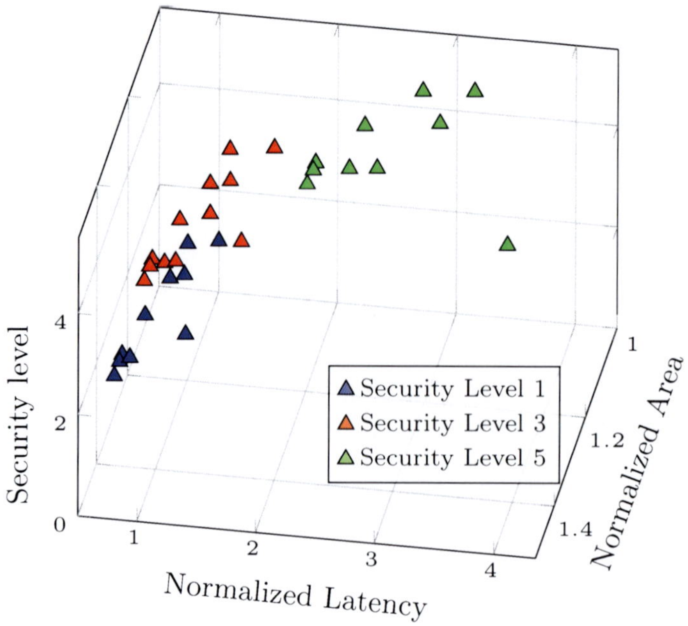

Fig. 9.1 Design-space exploration of key generation component of SPHINCS+ algorithm using Vivado-HLS directives

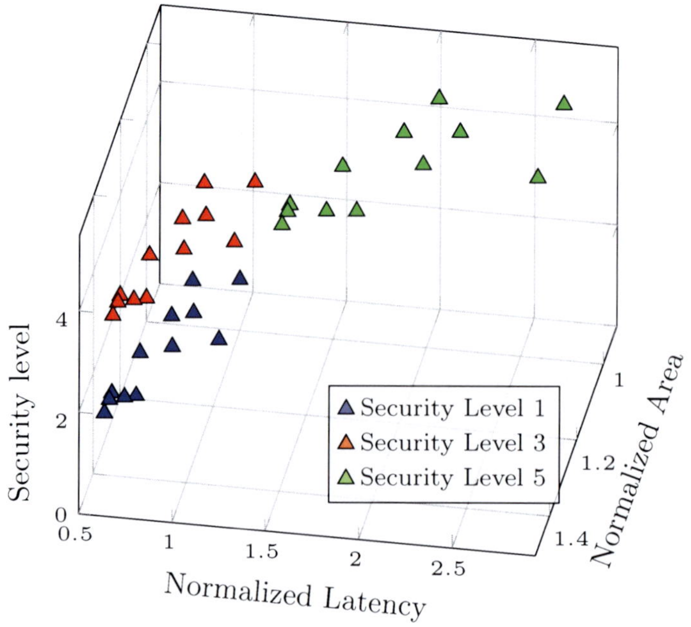

Fig. 9.2 Design-space exploration of SPHINCS+ signature generation algorithm using various Vivado-HLS directives

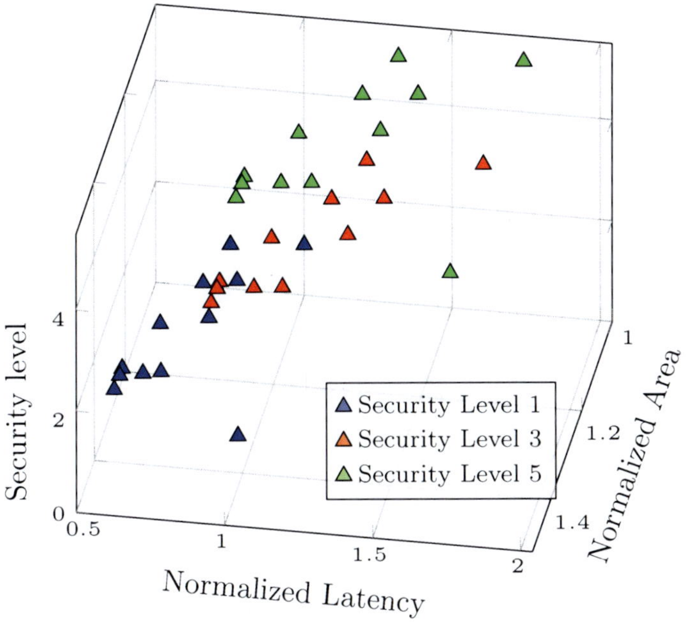

Fig. 9.3 Design-space exploration of SPHINCS$^+$ signature verification using Vivado-HLS directives

Takeaway The haraka hash is used to reduce the performance bottleneck within SPHINCS$^+$. For the sake of the FPGA implementation, we use haraka512.

9.4 ASIC-Specific Implementations and Optimizations

In this section, we discuss the ASIC implementation results. Tables 9.8, 9.9, and 9.10 report the power-performance-area trade-offs for key generation, signature generation, and signature verification, respectively.

As observed in Tables 9.8, 9.9, and 9.10, the loop-unrolled implementation provides a better throughput. Even though the clock period of the loop unrolled is higher, it has a better latency, hence a better execution time ($ClockPeriod \times$

Table 9.6 Description: Performance, area, and security (PAS) trade-off of SPHINCS$^+$ signature generation component for Artix-7 FPGA. Vivado-HLS directive loop unrolling and loop pipelining give various designs, which have different latency and area requirements, for three SPHINCS$^+$ variants (`sphincs-haraka-128f-simple`, `sphincs-haraka-192f-simple`, and `sphincs-haraka-256f-simple`)

Variant	Sec. level	Optimization	FF	LUT	BRAM	DSP	Clock (ns)	Latency *Kcycles*
`sphincs-haraka-128f-simple`	1	Base	21,183	78,671	83	0	7.097	1,234,532
		Unroll	28,569	107,903	149	0	7.727	665,887
		Pipe	21,187	79,782	86	0	7.097	934,227
`sphincs-haraka-192f-simple`	3	Base	21,241	79,187	84	0	7.966	1,339,292
		Unroll	28,659	108,511	148	0	7.966	725,730
		Pipe	21,245	80,298	87	0	7.966	1,014,830
`sphincs-haraka-256f-simple`	5	Base	20,775	78,243	91	0	7.097	3,359,530
		Unroll	28,161	107,484	157	0	7.727	1,834,593
		Pipe	20,779	79,354	94	0	7.097	2,553,864

Table 9.7 Description: Performance, area, and security (PAS) trade-off of SPHINCS$^+$ signature verification component for Artix-7 FPGA. Vivado-HLS directive loop unrolling and loop pipelining give various designs, which have different latency and area requirements, for three SPHINCS$^+$ variants (`sphincs-haraka-128f-simple`, `sphincs-haraka-192f-simple`, and `sphincs-haraka-256f-simple`).

Variant	Sec. level	Optimization	FF	LUT	BRAM	DSP	Clock (ns)	Latency *Kcycles*
`sphincs-haraka-128f-simple`	1	Base	15,160	57,847	58	0	8.651	50,562
		Unroll	20,326	78,196	104	0	8.651	27,431
		Pipe	15,164	58,615	60	0	8.651	38,333
`sphincs-haraka-192f-simple`	3	Base	15,398	58,554	59	0	8.651	81,546
		Unroll	20,580	78,967	104	0	8.651	44,430
		Pipe	15,402	59,322	61	0	8.651	61,917
`sphincs-haraka-256f-simple`	5	Base	14,997	57,853	67	0	8.651	88,118
		Unroll	20,163	78,211	113	0	8.651	48,277
		Pipe	15,001	58,621	69	0	8.651	67,061

Table 9.8 Description: ASIC power, performance, and area (PPA) for the different security levels of SPHINCS$^+$ key generation algorithm. Gate Count is technology independent and is derived by computing $(TotalArea - MemoryArea)\ /\ (Nand2\ gate\ area)$. The Nand2 gate used to obtain the above results has an area of 2.1 μm^2 with GF 65 nm LPE process technology. Given a gate count and a memory size, the ASIC area can be estimated in any technology. FF Memory size in kilobytes gives the amount of memory implemented using Flip Flops instead of SRAM or SROM. Execution time can be derived by computing $(ClockPeriod \times Latency)$

Variant	Sec. level	Optimization	Power mW	Clock period ns	Area in μm^2 SRAM/ ROM	Total area	Gate count $Kgate$	No of flip flops	Memory size in $Kbytes$ SRAM	SROM	FF	Latency $Kcycles$
`sphincs-`	1	Base	5.01	3.67	169,472	437,485	127.63	22,428	2.80	4.62	1.91	34,364
`haraka-`		Unroll	6.14	4.91	275,333	583,770	146.87	23,876	2.80	9.12	1.81	18,507
`128f-simple`		Pipe	5.06	3.72	169,472	440,564	129.09	22,453	2.80	4.62	1.91	25,969
`sphincs-`	3	Base	5.09	3.67	191,109	466,833	131.30	22,929	4.74	4.62	1.96	50,795
`haraka-`		Unroll	6.19	4.92	296,970	610,792	149.44	24,377	4.74	9.12	1.87	27,453
`192f-simple`		Pipe	5.19	3.85	191,109	469,949	132.78	22,954	4.74	4.62	1.96	38,438
`sphincs-`	5	Base	5.07	3.69	205,335	477,357	129.53	22,976	7.44	4.62	2.00	135,476
`haraka-`		Unroll	6.18	4.93	311,196	626,225	150.01	24,424	7.44	9.12	1.91	73,455
`256f-simple`		Pipe	5.19	3.74	205,335	481,797	131.65	23,001	7.44	4.62	2.00	102,646

Table 9.9 Description: ASIC power, performance, and area (PPA) of different security levels of SPHINCS$^+$ signature generation algorithms. Gate Count is technology independent and is derived by computing $(TotalArea - MemoryArea)$ / $(Nand2\ gate\ area)$. The Nand2 gate used to obtain the above results has an area of 2.1 μm^2 with GF 65 nm LPE process technology. Given a gate count and a memory size, the ASIC area can be estimated in any technology. FF Memory size in kilobytes gives the amount of memory implemented using Flip Flops instead of SRAM or SROM. Execution time can be derived by computing $(ClockPeriod \times Latency)$

Variant	Sec. level	Optimization	Power mW	Clock period ns	Area in μm^2 SRAM/ROM	Area in μm^2 Total area	Gate count $Kgate$	No of flip flops	Memory size in $Kbytes$ SRAM	SROM	FF	Latency $Kcycles$
sphincs-haraka-128f-simple	1	Base	13.24	3.98	280,637	1,072,180	376.93	65,283	4.42	8.12	5.50	1,234,532
		Unroll	16.51	4.95	633,506	1,548,437	435.68	69,929	4.42	23.12	5.18	665,887
		Pipe	13.45	4.06	280,637	1,081,928	381.57	65,362	4.42	8.12	5.50	934,227
sphincs-haraka-192f-simple	3	Base	13.28	3.95	319,842	1,120,599	381.31	66,315	7.97	8.12	5.60	1,339,292
		Unroll	16.54	4.94	672,712	1,595,261	439.31	70,961	7.97	23.12	5.29	725,730
		Pipe	13.49	4.11	319,842	1,132,148	386.81	66,394	7.97	8.12	5.60	1,014,830
sphincs-haraka-256f-simple	5	Base	13.22	3.92	352,685	1,140,847	375.32	65,005	12.97	8.12	5.50	3,359,530
		Unroll	16.57	4.94	705,555	1,613,658	432.43	69,651	12.97	23.12	5.18	1,834,593
		Pipe	13.44	3.89	352,685	1,149,249	379.32	65,084	12.97	8.12	5.50	2,553,864

Table 9.10 Description: ASIC power, performance, and area (PPA) of different security levels of SPHINCS+ signature verification algorithms. Gate Count is technology independent and is derived by computing $(TotalArea - MemoryArea)$ / $(Nand2\ gate\ area)$. The Nand2 gate used to obtain the above results has an area of 2.1 μm^2 with GF 65 nm LPE process technology. Given a gate count and a memory size, the ASIC area can be estimated in any technology. FF Memory size in Kbytes gives the amount of memory implemented using Flip Flops instead of SRAM or SROM. Execution time can be derived by computing $(ClockPeriod \times Latency)$

Variant	Sec. level	Optimization	Power mW	Clock period ns	Area in μm^2 SRAM/ROM	Total area	Gate count $Kgate$	No of flip flops	Memory size in $Kbytes$ SRAM	SROM	FF	Latency $Kcycles$
sphincs-haraka-128f-simple	1	Base	9.84	4.05	199,434	808,114	289.85	51,099	6.27	6.91	4.47	50,562
		Unroll	11.49	4.97	446,443	1,131,940	326.43	54,345	6.27	17.41	4.25	27,431
		Pipe	9.98	4.04	199,434	814,095	292.70	51,153	6.27	6.91	4.47	38,333
sphincs-haraka-192f-simple	3	Base	9.99	3.81	267,489	884,891	294.00	52,143	12.41	9.80	4.57	81,546
		Unroll	11.80	4.92	514,498	1,222,595	337.19	55,389	12.41	20.30	4.35	44,430
		Pipe	10.02	3.82	267,489	891,152	296.98	52,197	12.41	9.80	4.57	61,917
sphincs-haraka-256f-simple	5	Base	9.85	3.92	311,634	915,444	287.53	50,945	20.66	13.56	4.46	88,118
		Unroll	11.68	4.91	558,642	1,255,164	331.68	54,191	20.66	24.06	4.24	48,277
		Pipe	10.00	4.03	311,634	921,832	290.57	50,999	20.66	13.56	4.46	67,061

Latency). However, the loop-unrolled version presents the highest area and power. On the other hand, the baseline version provides better power, area, and clock period, but its latency is very high, leading to a higher execution time. Moreover, for all the three variants and a given optimization technique, the standard cell area (shown in the Gate Count column) is almost the same for all variants, whereas memory requirements (SRAM column) are different: SROM size for the loop unrolled version is higher compared to the other two optimizations. In Table 9.9, the memory, as implemented with Flip Flops (FF column), is quite significant at ~5 KB, because there are around 200 small-size memory constructs, for which conversion to SROM/SRAM does not provide any area benefits.

Table 9.11 reports a high area requirement for internal SPHINCS$^+$ modules for key generation, signature generation, and signature verification, respectively. Based on these reports, key design blocks in SPHINCS$^+$ are `treehash_wots`, `randombytes`, `tweak_constants`, `AES256_CTR_DRBG_Update`, `fors_sign`, `hash_message`, `wots_sign`, `fors_pk_from_sig`, `compute_root`, `wots_pk_from_sig`, and `thash_3`.

Table 9.11 Description: area consumed by the design blocks used in SPHINCS$^+$

	% of total area								
	sphincs-haraka-128f-simple			sphincs-haraka-192f-simple			sphincs-haraka-256f-simple		
Design blocks	Base	Unroll	Pipe	Base	Unroll	Pipe	Base	Unroll	Pipe
Key generation									
treehash_wots	48.9	53.9	49.3	51.5	55.6	51.8	52.4	56.3	52.8
randombytes	25.8	19.3	25.6	24.2	18.5	24.1	23.6	18.0	23.4
tweak_constants	23.2	25.2	23.0	21.8	24.1	21.7	21.2	23.6	21.1
AES256_CTR_DRBG_Upda	13.1	9.8	13.0	12.3	9.4	12.2	12.0	9.2	11.9
Signature generation									
fors_sign	24.7	24.8	24.7	25.4	25.3	25.5	24.8	24.9	24.8
treehash_wots	20.3	20.5	20.4	21.3	21.2	21.5	22.3	21.9	22.4
hash_message	14.7	16.0	14.6	14.1	15.6	14.0	13.9	15.4	13.8
randombytes	10.6	7.3	10.5	10.1	7.1	10.0	9.9	7.0	9.8
wots_sign	8.2	9.2	8.4	8.1	9.0	8.3	8.6	9.4	8.8
tweak_constants	5.7	5.4	5.7	5.5	5.3	5.4	5.4	5.2	5.4
Signature verification									
fors_pk_from_sig	37.4	37.1	37.4	37.8	37.4	37.9	37.8	37.5	37.9
compute_root	21.1	20.8	20.9	21.9	21.4	21.8	21.8	21.4	21.7
hash_message	15.0	16.5	14.9	13.7	15.4	13.6	13.2	15.0	13.1
wots_pk_from_sig	7.9	8.4	8.2	7.5	8.1	7.8	8.0	8.5	8.3
tweak_constants	7.5	7.3	7.4	6.8	6.8	6.8	6.6	6.6	6.6
thash_3	6.9	6.8	6.8	7.1	7.0	7.0	7.0	7.0	7.0

`treehash_wots` uses Merkle's tree hash algorithm to compute the root node and authentication path for an input leaf. `randombytes` generates a pseudo-random number using AES. `tweak_constants` is part of the haraka hash function. It changes the AES key that is used to generate collision-resistant hash values. `AES256_CTR_DRBG_Upda` (or `AES256_CTR_DRBG_Update`) performs AES in the counter mode. `fors_sign` extracts the secret key from `sk_seed` and signs the message using FORS. hash_message computes hash values using haraka, with inputs as a randomized value, the public key, and the message. `wots_sign` calculates a signature with an n-byte message and a secret key seed. `fors_pk_from_sig` extracts FORS public key from the signature. `compute_root` takes the authentication path and a leaf to calculate a root node. `wots_pk_from_sig` extracts WOTS public key from a WOTS signature and an n-byte message. `thash_3` generates the hash value for the input array.

Takeaway Memory requirements vary across the three optimizations, but there are minimal changes in the standard cell area.

9.5 Key Takeaways

- SPHINCS$^+$ components have a clock period of 8.651 ns. yielding a maximum frequency of ~115 MHz.
- In terms of area requirements, the order is as follows: key generation < signature verification < signature generation. In terms of execution latency: key generation < signature verification < signature generation.
- In the case where DSPs are not included in the area overhead, we make the following observations: area of security level 5 < area of security level 1 < area of security level 3. Security level 5 provides the highest security with minimum area overhead. However, this incurs a large latency in comparison to other security levels.
- Among the three components, signature verification, with a large combinational logic, requires a larger clock period.
- Latency requirement of SPHINCS$^+$ components loosely follows the trend:

$$(SPHINCS^+ \ level\text{-}1) : (SPHINCS^+ \ level\text{-}3) : (SPHINCS^+ \ level\text{-}5) = 1 : 2 : 3$$

- SPHINCS$^+$ does not require any additional memory elements for temporary storage.
- For key generation, loop unrolling optimizations improve the latency by ~50%, increase LUTs by ~33%, and increase FFs by ~50%.
- Loop pipelining improves the latency by 18% with slight incurred area overhead.
- Figure 9.1 shows several design points that improve latency and increase the required area. It shows significant latency improvement using optimizations.
- The design space exploration exhibits the fundamental area-speed trade-off curve. Depending on the requirements, the design point is chosen accordingly.
- Security level 5 incurs the highest latency. Figure 9.1 reveals that the fastest level of security level 5 design, using parallel hardware incurs more latency than the slower designs of security level 1 or security level 3. In faster applications, security level 1 or security level 3 are better choices, if the security requirements are met.
- Signature generation takes 20× latency/time requirements as compared to signature verification and key generation.
- For signature generation, each variant occupies a similar area overhead on FPGA implementations. However, security level 5 takes ~10% more BRAMs than other security levels.
- Signature generation incurs more latency than the other two components. Hence, performance optimization on this component is critical for faster applications. Latency for security level 5 is ~3× more than that of security level 1 and security level 3. Security level 1 is better suited for faster implementations.
- As shown in Table 9.6, loop unrolling improves the latency by ~50% for a ~25% area increase compared to the baseline.
- As reported in Table 9.6, loop pipelining improves the latency by ~25% with minimal area overhead. As latency numbers are quite large for signature generation, loop unrolling provides better optimization for this component.
- Figure 9.2 shows that optimizations improve the latency significantly for a small area increase. Security level 5 requires more latency even for the fastest design.
- Figure 9.2 shows that security levels 1 and 3 have similar area and latency requirements. Hence, security level 3 is a better candidate with a higher security strength.
- For signature verification, the latency of security level 5 is ~80% more than security level 1's latency and security level 3 is ~60% more than security level 1.
- As shown in Table 9.7, loop unrolling improves the latency by ~50% for a ~25% increase in area compared to the baseline case.
- As reported in Table 9.7, loop pipelining improves latency by ~25% with little area overhead.
- Figure 9.3 shows that security level 3 and security level 5 have similar area and latency requirements. Optimized security level 1 designs, however, require ~40% less latency. Security level 1 is the fastest design.

- Figure 9.3 reveals that different SPHINCS$^+$ signature verification architectures do not have significant differences in area or latency. Optimized security level 5 is a better candidate as it provides better speed, area, and security, compared to the other architectures.
- The number of flip flops is more in ASIC than the FPGA implementation. This is because all the memory is implemented using BRAM in FPGA. But in ASIC, the memory is converted to SRAM only if the conversion gives area gains; thus, smaller memory is not implemented as SRAM, but with flip flops.
- Figures 9.4, 9.5, and 9.6 show that it is possible to select the right optimization technique for security level 3, which gives comparable PPA with security level 1.
- When compared to 28 nm FPGA frequency numbers, ASIC in 65 nm technology is around 55% faster for key generation and signature generation algorithms and 75% faster for the signature verification algorithm.
- With evaluation and exploration SPHINCS$^+$ hardware designs, we completed exploration of eight NIST PQC round-2 signature schemes.

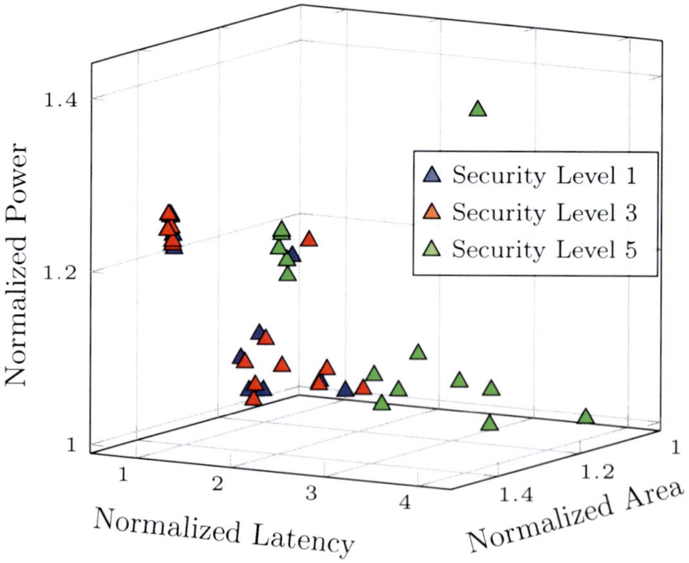

Fig. 9.4 ASIC design-space exploration of SPHINCS$^+$ key generation component when implemented using loop unrolling and pipelined optimizations, normalized to a baseline of security level 1 key generation

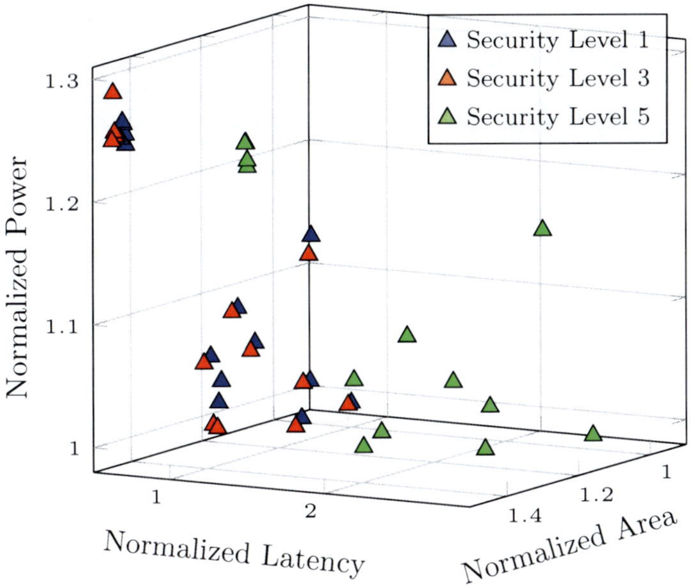

Fig. 9.5 ASIC design-space exploration of SPHINCS$^+$ signature generation component when implemented using loop unrolling and pipelined optimizations, normalized to a baseline of security level 1 signature generation

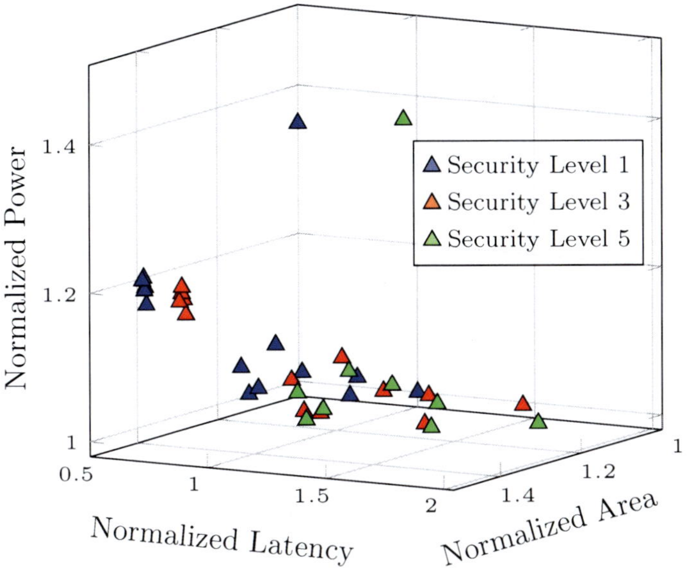

Fig. 9.6 ASIC design-space exploration of SPHINCS$^+$ signature verification component when implemented using loop unrolling and pipelined optimizations, normalized to a baseline of security level 1 signature verification

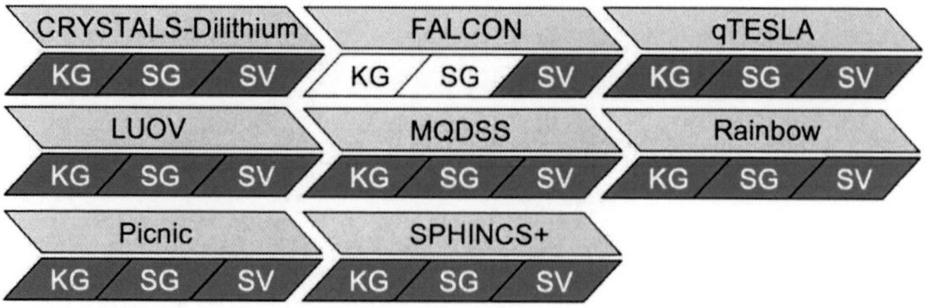

References

1. A. Hulsing, D.J. Bernstein, C. Dobraunig, M. Eichlseder, S. Fluhrer, S.-L. Gazdag, P. Kampanakis, S. Kolbl, T. Lange, M.M. Lauridsen, F. Mendel, R. Niederhagen, C. Rechberger J. Rijneveld, P. Schwabe, J.-P. Aumasson, *Sphincs+ Submission to the NIST Post-Quantum Cryptography Standardization Project* (2019). https://csrc.nist.gov/CSRC/media/Projects/Post-Quantum-Cryptography/documents/round-2/submissions/SPHINCS-Round2.zip
2. D.J. Bernstein, A. Hülsing, S. Kölbl, R. Niederhagen, J. Rijneveld, P. Schwabe, The sphincs+ signature framework, in *Proceedings of the 2019 ACM SIGSAC Conference on Computer and Communications Security*, 2019, pp. 2129–2146
3. D.J. Bernstein, D. Hopwood, A. Hülsing, T. Lange, R. Niederhagen, L. Papachristodoulou, M. Schneider P. Schwabe, Z. Wilcox-O'Hearn, Sphincs: practical stateless hash-based signatures, in *Annual International Conference on the Theory and Applications of Cryptographic Techniques*, 2015, pp. 368–397
4. A. Hülsing, J. Rijneveld, F. Song, Mitigating multi-target attacks in hash-based signatures, in *Public-Key Cryptography—PKC 2016*, 2016, pp. 387–416
5. J.-P. Aumasson, G. Endignoux, Improving stateless hash-based signatures, in *Cryptographers' Track at the RSA Conference* 2018, pp. 219–242
6. D.J. Bernstein, A. Hülsing, Decisional second-preimage resistance: When does SPR imply PRE? in *International Conference on the Theory and Application of Cryptology and Information Security* (Springer, Berlin, 2019), pp. 33–62
7. S. Kölbl, M.M. Lauridsen, F. Mendel, C. Rechberger, Haraka v2–efficient short-input hashing for post-quantum applications, in *IACR Transactions on Symmetric Cryptology*, 2016, pp. 1–29

Chapter 10
Conclusion

10.1 Goals of This Study

The goals of this book are twofold as clearly articulated by the forewords.

Goal 1: The NIST cryptography expert articulates that, "a crucial goal during this transition period is the scientific examination and benchmarking of the performance of candidate-algorithms in various protocols, platforms, and types of hardware. In this book, the authors present the first comprehensive suite of experiments for hardware designs of NIST's post-quantum candidate-algorithms. The approach taken is the use of a High-Level Synthesis (HLS) tool to produce hardware implementations on FPGAs and ASICs. This is an exciting and important first step into the expansive world of post-quantum cryptographic hardware, which will drive global communications and commerce for future decades."

Goal 2: The Xilinx HLS expert states that "this work represents a key first step to begin exploring the implementation space of new post-quantum cryptographic algorithms. By adapting the reference processor-oriented code to HLS-friendly code, the authors have demonstrated the feasibility of hardware implementations of these algorithms and provided a baseline to compare other hardware implementations too. The path is clear for wider analysis of cost vs. performance trade-offs and optimization of these implementations in concert with analysis of the algorithms themselves."

Consistent with these goals, in this book, we develop and apply a uniform hardware design methodology that enables a common set of optimizations for implementing NIST PQC algorithms in hardware and allowing for a common benchmarking method. Given the complexity of the algorithms, generating functionally correct RTL designs in Verilog/VHDL under standardization driven time constraints is challenging. We develop and adopt an HLS-based approach starting with the C specifications submitted to the NIST evaluation process. A similar

D. Soni et al., *Hardware Architectures for Post-Quantum Digital Signature Schemes*, https://doi.org/10.1007/978-3-030-57682-0_10

HLS-based design flow has been successfully adopted to evaluate cryptographic algorithms submitted to the CAESAR competition [1]. While the HLS-based approach cannot replace a hand-optimized RTL code, we deem it prudent to use it for early and rapid analysis of hardware trade-offs. With this objective, we presented preliminary FPGA-based hardware evaluation of NIST PQC algorithms in [2, 3].

In general, HLS-based design space exploration significantly reduces overall development time when compared to handwritten and optimized RTL designs. Hence, HLS is being increasingly used by the industry and academia as a faster, easier, and less error-prone method for hardware design. Prior studies confirm that HLS-based approaches yield results comparable to those obtained using handcrafted RTL for traditional cryptography [4]. Both approaches deliver similar results in terms of area and frequency, while the HLS-based approach has less throughput/performance. The HLS generated RTL code can benchmark power, area, and speed for hardware implementations of cryptography algorithms. If the resulting RTL does not meet these requirements, the designer can replace parts of it with handwritten RTL. The quality of results of HLS-generated FPGA or ASIC implementations is determined by the quality of the input C code. For instance, hardware-aware input C code will yield better final designs than a C code that is more processor-instruction-set-architecture-aware. In this study, we did not focus on this aspect of the design flow. However, this is an important future direction that the community can benefit from.

10.2 Discussion

The NIST PQC competition round-2 has shortlisted nine digital signature candidates. In this book, we have evaluated the relative performance of all these algorithms, except for GeMSS,[1] through a thorough design space exploration.

Table 10.1 shows a few of the key building blocks of PQC signature schemes that we have evaluated in this book. NTT is one of the key building blocks for lattice-based PQC signature schemes. Hash function and block cipher (AES) are part of all PQC signature schemes. Hash function has multiple variants that developers have used. For example, SPHINCS$^+$ uses haraka, SHA3, and SHAKE as hash function for different variants. Each of these variants has advantage/disadvantage depending on the underlying hardware or firmware. If these building blocks are optimized for HLS, it would improve performance and area utilization of all PQC signature schemes. For several signature schemes, we have performed directive optimization to improve area utilization of hash function. As these hash functions are used frequently, it showed significant area reduction.

[1]For signature verification of GeMSS, we generated a hardware design, but we have not completed design-space exploration as of October 2020.

Table 10.1 Building blocks for hardware design of signature schemes

Algorithm	AES	Hash function	NTT
CRYSTALS-Dilithium	✓	✓[a, b]	✓
Falcon	✓	✓[b]	✓
qTesla	✓	✓[a, b]	✓
LUOV	✓	✓[a, b]	
MQDSS	✓	✓[a, b]	
Rainbow	✓	✓[c]	
Picnic	✓	✓[a, b]	
SPHINCS$^+$	✓	✓[d]	

[a]SHAKE128, [b]SHAKE256, [c]SHA2-256/-384/-512, and [d]haraka256/512

Table 1 contains the implemented PQC signature schemes along with the links to the HLS-ready C code for each of these algorithms. These online resources contain the relevant synthesis directives and Vivado HLS project.

10.3 Contribution to the Research Community

This book is intended to benefit the hardware and PQC research community:

1. HLS-based design space exploration enables rapid benchmarking of PQC accelerators. With further exploration informed by profiling, designers can identify resource-intensive modules and perform additional, targeted optimizations. For example, a designer can identify computationally intensive operations such as modular reduction, the number theoretic transform (NTT), Inverse NTT, hash functions, and block ciphers and perform targeted optimizations on these operations (or replace them with optimized modules such as for AES block cipher, SHA-3 and other hash functions). HLS-based design enables rapid profiling of the PQC algorithm performance as a function of the security levels and as the schemes evolve during NIST standardization.
2. Designers can use the HLS-generated FPGA and ASIC designs and their components (e.g., the datapath, the control logic, memory, and memory control logic) and their performance characteristics (e.g., timing, area, and power) to assess the side-channel information leakage and the different types side-channel vulnerabilities. This can in turn inform integration of countermeasures against one or more of the side-channel attacks.
3. Instructors can use the HLS-synthesizable C code from the repository in Table 1 in their classes or as starting points for basic and advanced projects. These C codes can be targeted to FPGAs platforms or ASIC technologies. Designers can hone their HLS expertise using PQC as the application domain.

10.4 Status Update

As of June 2020, the NIST PQC standardization process is not complete. The PQC signature schemes and the parameters are subject to change. We will update the hardware designs of these schemes and will make the results available at https://wp.nyu.edu/hipqccheck/book-pqc-signature-schemes/. On this page, we will update the hardware evaluation and optimizations, and answer commonly asked HLS-related design questions. The Technology Innovation Institute team, which is conducting the cryptographic analysis of the presented digital signature schemes, will answer related questions.

References

1. E. Homsirikamol, K.G. George, Toward a new HLS-based methodology for FPGA benchmarking of candidates in cryptographic competitions: the CAESAR contest case study, in *IEEE International Conference on Field Programmable Technology*, 2017, pp. 120–127
2. K. Basu, D. Soni, M. Nabeel, R. Karri, *NIST Post-Quantum Cryptography: A Hardware Evaluation Study*, IACR Cryptology ePrint Archive, Report 2019/047 (2019). https://eprint.iacr.org/2019/047
3. D. Soni, K. Basu, M. Nabeel, R. Karri, A hardware evaluation study of NIST post-quantum cryptographic signature schemes in *Second PQC Standardization Conference* (NIST, 2019). https://csrc.nist.gov/CSRC/media/Events/Second-PQC-Standardization-Conference/documents/acceptedpapers/sonihardwareevaluation.pdf
4. E. Homsirikamol, K. Gaj, Can high-level synthesis compete against a hand-written code in the cryptographic domain? A case study, in *International Conference on ReConFigurable Computing and FPGAs*, 2014, pp. 1–8

Glossary

Digest is the output of a cryptographic hash function.

Existential unforgeability under chosen-message attack (EUF-CMA) In this attack, adversary can create (forge) at least one message/signature pair (m, σ) where message (m) was not signed by authorized user in the past.

High-level synthesis is an automated process that converts a high-level specification of an algorithm to hardware description language.

Hash function/cryptographic hash function One-way function to map input data of any size in fixed size output data.

Known answer tests (KATs) KATs are reference input–output test data for each PQC candidate. They produce results for fixed testcases provided by NIST and contain a seed, message, message length, public key, secret key, signature, and signature length.

Key generation Digital signature component which generates a pair of keys (secret key and public key). It is required by any digital signature when the keys are ephemeral or varying.

Multi-party computation (MPC) is an encryption protocol that jointly encrypts or signs the message with different servers, processors, or computational units. An individual computational unit does not know the complete value of the message or key.

Nonce is a non-repeating arbitrary number needed only once in a cryptographic protocol.

Permutation A process to rearrange elements of a set or sequence or array.

Post-quantum cryptography Cryptographic algorithms that are secure against an attack by classical computers or quantum computers.

Pseudo-random number generator An algorithm that generates random looking output that is not truly random, because it is calculated in a deterministic manner from an initial value (seed).

Public key Key used to verify the authenticity of the message.

D. Soni et al., *Hardware Architectures for Post-Quantum Digital Signature Schemes*, https://doi.org/10.1007/978-3-030-57682-0

Public key infrastructure (PKI) is a framework that manages digital certificates and public-key encryption to authenticate devices and users in the digital world.

Salt Random input data, along with an actual hash input, for any one-way function. It works as an additional safeguard for a one-way function, especially a hash function.

Side-channel attack Any attack which extracts secrets based on the analysis of information collected from the implementation of a system/algorithm.ă

Secret key/private key Key used to encrypt/sign the original message.

Security level NIST has classified the security strength of Post-Quantum Cryptographic algorithms in five categories; Security level 1 (the least) to Security level 5 (strongest).

Secure hash function (SHA) A family of cryptographic hash functions design by NIST to keep data secure.

Signature/signed message It is the sender's signed message generated by appending the original message and a cryptographic version of the message which are encrypted with a secret key.

Signature generation Digital signature component that encrypts the original message using a public key and outputs a signature or signed message.

Signature verification Digital signature component that verifies a digital signature on data/a message using a public key.

Strong unforgeability under chosen message attacks (SUF-CMA) In this attack, adversary has a message/signature pair (m, σ). Based on this, different signature (σ') is created for same message (m).

Index

D. Soni et al., *Hardware Architectures for Post-Quantum Digital Signature Schemes*, https://doi.org/10.1007/978-3-030-57682-0

MIX
Papier aus verantwortungsvollen Quellen
Paper from responsible sources
FSC® C105338

Printed by Books on Demand, Germany